Make: Volume 26

40: Red Racer Drill Kart
Build a zippy go-kart powered by twin cordless drills.
By Gever Tulley & MAKE Labs

46: World's Simplest Longboard
Make a terrific skateboard with a minimum of fuss or tools.
By Mark Frauenfelder

48: Make a Motorbike
Add a retro-style, one-cylinder gasoline engine to your bike.
By Lew Frauenfelder

50: OutRun Arcade Car
Building the un-simulation of a driving video game.
By Garnet Hertz

54: Kult of the Kart
Cyclekarts are tiny racers optimized for fun. By Nik Schulz

56: Faster Than the Wind
Mind-bending propeller cart outruns the wind that powers it.
By Eric Chu

60: Gravity Racer
Race past the competition with the Make: Projects contest winner: an all-metal, quad-suspension soapbox cart.
By Jeremy Ashinghurst

65: Wild Wheeled World
A roundup of DIY karts, scooters, and other rolling wonders.

Columns

11: Welcome: That's How We Roll
Karts bring out the kid in us all. By Mark Frauenfelder

12: Reader Input
Arduinophytes and maker services.

15: Maker's Calendar
Events from around the world. By William Gurstelle

16: In the Maker Shed
Get projects in your mailbox. By Dan Woods

26: Tales from Make: Online
Live from MAKE! By Gareth Branwyn

27: Making Trouble
DIT: Raising our collective barn. By Saul Griffith

28: Country Scientist
Ultra-simple sunshine recorders. By Forrest M. Mims III

31: Make Free: Moral Suasion
A wake-up call for anyone who's blithely relying on the cloud.
By Cory Doctorow

RED RACER: Anyone can build Gever Tulley's nifty go-kart, which is powered by a pair of cordless drills. Photography by Garry McLeod. Prop/wardrobe styling by Tietjen Fischer. Makeup/hair styling by Renee Rael.

SIMPLE SKATE: Make this smooth-riding longboard in about a day using plywood and glue.

WEEKEND WARRIOR: Ashinghurst's gravity racer won first place in our Make: Projects contest.

Vol. 26, April 2011. MAKE (ISSN 1556-2336) is published quarterly by O'Reilly Media, Inc. in the months of January, April, July, and October. O'Reilly Media is located at 1005 Gravenstein Hwy. North, Sebastopol, CA 95472, (707) 827-7000. SUBSCRIPTIONS: Send all subscription requests to MAKE, P.O. Box 17046, North Hollywood, CA 91615-9588 or subscribe online at makezine.com/offer or via phone at (866) 289-8847 (U.S. and Canada); all other countries call (818) 487-2037. Subscriptions are available for $34.95 for 1 year (4 quarterly issues) in the United States; in Canada: $39.95 USD; all other countries: $49.95 USD. Periodicals Postage Paid at Sebastopol, CA, and at additional mailing offices. POSTMASTER: Send address changes to MAKE, P.O. Box 17046, North Hollywood, CA 91615-9588. Canada Post Publications Mail Agreement Number 41129568. CANADA POSTMASTER: Send address changes to: O'Reilly Media, PO Box 456, Niagara Falls, ON L2E 6V2

Make: Projects

Sonic Flame Tube

See the shape of sound, written in tongues of fire. By William Gurstelle

72

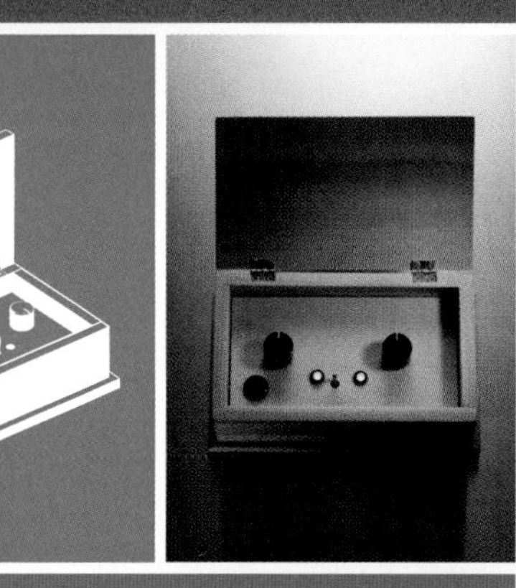

Luna Mod

Make a simple handheld synthesizer and looper box to generate intriguing sonic rhythms. By Brian McNamara

80

Spirulina Superfood Tank

Supplement your daily diet with fresh spirulina "superfood" grown indoors next to a sunny window. By Aaron Wolf Baum

92

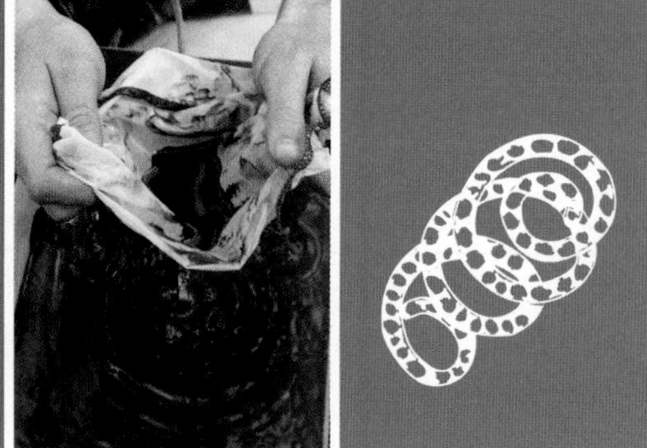

PRIMER

BIOSENSING

Track your body's signals and brain waves, and use them to control things.
By Sean M. Montgomery and Ira M. Laefsky

102

Make: Volume 26

Maker

18: **Made on Earth**
Snapshots from the world of backyard technology.

32: **Manufacture Your Project**
Take your invention from concept to mass production and live the dream. By Mitch Altman

38: **1+2+3: Paper Clip Paper Holder**
A giant clip for your desktop. By Gus Dassios

70: **1+2+3: Bokeh Photography Effect**
Create an effect that makes the out-of-focus lights in your pictures appear any shape you want. By Sindri Diego

112: **1+2+3: Jumper Wires**
An easy way to make protoboard wires. By Charles Platt

142: **Electronics: Fun and Fundamentals**
Hack an alarm siren to make a DIY dog repeller. By Charles Platt

146: **Make Money**
Combination lock made from coins. By Tom Parker

148: **Toys, Tricks, and Teasers**
Deceive the eye with kinetic toy illusions. By Donald Simanek

152: **MakeShift: Charge That Cell!**
Stuck in the woods with dead batteries. By Lee D. Zlotoff

154: **Howtoons: Spool Racer**
By Saul Griffith and Nick Dragotta

156: **Toolbox**
The best tools, gadgets, and resources for makers.

164: **Toy Inventor's Notebook**
Make a diving SpudMarine! By Bob Knetzger

166: **What I Made**
Fashion a bird feeder from soda cans and chopsticks. By Scott Bedford

168: **Heirloom Technology: Chinampa Gardens**
Craft a self-watering planter using Aztec agricultural techniques. By Tim Anderson

170: **Remaking History: Bag Press**
Use linen, wood, and the wisdom of ancient Egyptians to make a juicer. By William Gurstelle

176: **Homebrew: My Power Wheelchair**
By Marcus Brooks

Family fun project

READ ME: Always check the URL associated with a project before you get started. There may be important updates or corrections.

32

MAKER MADE: TV-B-Gone inventor Mitch Altman says manufacturing your invention is its own reward.

DIY

113: Circuits
Pendulum challenge, pushbutton hacks.

120: Outdoors
Solar food dryer, fabric-formed posts.

128: Home
Fool's stool, jigsaw puzzle chair, magnetic knife holder, simple art easel.

148

LOOPY LOGIC: Tumbling rings are a spectacular illusion you can construct from key rings.

> "True happiness comes from the joy of deeds well done, the zest of creating things new."
> —*Antoine de Saint-Exupéry*

PUBLISHED BY
O'REILLY MEDIA, INC.
Tim O'Reilly, CEO
Laura Baldwin, COO

Visit us online:
makezine.com

CUSTOMER SERVICE
cs@readerservices.makezine.com

Manage your account online, including change of address:
makezine.com/account
866-289-8847 toll-free in U.S. and Canada
818-487-2037, 5 a.m.–5 p.m., PST

Follow us online:
On Twitter: @make
On Facebook: makemagazine

Comments may be sent to:
editor@makezine.com

MAKER MEDIA CARES
MAKE is printed on recycled, process-chlorine-free, acid-free paper with 30% post-consumer waste, certified by the Forest Stewardship Council and the Sustainable Forest Initiative, with soy-based inks containing 22%–26% renewable raw materials.

PRINTED WITH SOY INK

PLEASE NOTE: Technology, the laws, and limitations imposed by manufacturers and content owners are constantly changing. Thus, some of the projects described may not work, may be inconsistent with current laws or user agreements, or may damage or adversely affect some equipment.

Your safety is your own responsibility, including proper use of equipment and safety gear, and determining whether you have adequate skill and experience. Power tools, electricity, and other resources used for these projects are dangerous, unless used properly and with adequate precautions, including safety gear. Some illustrative photos do not depict safety precautions or equipment, in order to show the project steps more clearly. These projects are not intended for use by children.

Use of the instructions and suggestions in MAKE is at your own risk. O'Reilly Media, Inc., disclaims all responsibility for any resulting damage, injury, or expense. It is your responsibility to make sure that your activities comply with applicable laws, including copyright.

Contributors

William Gurstelle's (*Flame Tube*) neighbors used to wonder just what was going on in his Minneapolis backyard. It took a while, but they seem to have become acclimated to the occasional catapult firings, potato cannon booms, and smoke from the rocket launches. The flamethrower blasts? Well, still not so much. As one of MAKE's contributing editors and technical advisory board members, Bill has written scores of articles for this magazine. His newest book, *The Practical Pyromaniac*, reminds a lot of people of his first book, *Backyard Ballistics*, and not just because they're both alliterative titles. thepracticalpyromaniac.com

Liz Llewellyn's (*Simple Art Easel*) first sentence was "I'll do it myself!" As a teenager, she designed and made a sweater from scratch, even inventing her own crochet stitch. As an adult, she studied mechanical engineering and decided to "seek out a career that would be best described by the sentence, 'I make cool things that work.' Which is what I do now, and it's even cooler than I imagined it would be." She lives in a 100-year-old house in an old mill town west of Boston with her husband, one teenager, one preschooler, two dogs, and a cat. What's her favorite thing to make? "The thing I haven't made yet."

Juan Leguizamon (Special Section opening illustration) lives in the TenderNob district of San Francisco, and has been "tracing" since he was 5. "Now when I see something that I really like, I mentally trace it in my head and make it into something cool on paper," he says. He's a senior designer at Razorfish, an interactive ad agency, does freelance illustration, is "tempted to have a cat even though I've always been more of a dog person," and has always been fascinated with cardboard. "The idea of making something that is not meant to be final or finished intrigues me. It's just like when we were little and with our imagination we could make a simple cardboard box be anything we wanted."

Linda Nguyen's (*Simple Longboard* photography) natural curiosity has led to many photographic adventures in her life. Whether she is out on her own shooting a personal project or on assignments with advertising and editorial clients such as Nike and Enterprise, she's fueled by the inspiration of the given moment. Her current projects include making a music video and concocting the perfect pho recipe, complete with hand-drawn illustrations and instructions, to pass down to her friends and family. She is based in Southern California, where she enjoys the warmth of the sun, sips tea, and listens to jazz records.

Aaron Wolf Baum (*Spirulina Superfood Tank*), aka Dr. Friendly, came to his expertise in DIY algae superfood by way of Harvard, Stanford, Silicon Valley, and Burning Man. He has "a passion for sustainability, teaching, and science in general." Not only is he very friendly, he's committed to using his talents "to do something good, and right now that means looking hard at how our energy choices affect the natural world that sustains us." Check his website (drfriendly.tv) and his blogs (farmeronmars.blogspot.com and draaronwolfbaum.wordpress.com) to learn more about the doctor's pursuits.

Brian McNamara (*Luna Mod*) has been pulling things apart since he was 2 or 3 years old. Many years later he has not only learned how to put some of them back together, but is even able to make new things from scratch. After working in a wide range of electronics workshops, from avionic to scientific, Brian finally decided to set up his own workshop, designing and building unique electronic devices, mostly electronic musical instruments that he sells through his online business (rarebeasts.com, @rarebeasts). Brian loves making music, gardening, and taking his two kids on hiking adventures.

That's How We Roll

I WAS IN THE SECOND GRADE WHEN MY father built a go-kart for my cousin and me. It was made from lumber and four new wheelbarrow tires. Steering was accomplished by putting one foot on either end of the front axle, which was connected to the go-kart's frame with a single large bolt. To slow down while going downhill, all you had to do was grab the wooden lever attached to the side of the go-kart with another large bolt, pull back on it, and let friction do its work as the wood dug into the asphalt (we had to replace the brake frequently).

The go-kart was the hit of the neighborhood. All the kids wanted to ride it, and we'd take turns weaving around the tin-can slalom course we'd set up on a gently sloped road, timing our runs with a stopwatch. Soon other dads in the neighborhood built go-karts for their kids and we'd race each other.

In time we graduated from go-karts to minibikes. These gasoline-powered two-wheelers were a step up from gravity-powered vehicles. The noise, smell, and power of the minibikes intoxicated us for a few summers in our junior high school years.

Those memories came flooding back when we started creating the Karts and Wheels package in this issue of MAKE. I hope the projects in these pages will help you rekindle some of your own childhood memories, as well as provide your kids with opportunities for fun, adventure, and learning.

It probably goes without saying, but I'm going to say it anyway: Wear a helmet! When I was a kid, nobody wore helmets when they rode bikes, skateboards, go-karts, or minibikes. Please don't make that mistake today.

One spring day when I was 15 I rode my skateboard down a steep hill, and the only thing I remember from that afternoon is waking up covered in blood in the back of an

ambulance. I broke my nose, lost some teeth, received stitches for several cuts on my face, and concussed my brain. Fortunately, I didn't suffer anything worse. Today, my kids put on their helmets automatically whenever they ride their bikes or skateboards.

If go-karts and minibikes aren't your thing, don't despair — we have plenty of wonderful projects in this issue. We'll show you how to make a Rubens tube that emits jets of flame to display waveforms of any sound you play through it. We'll show you how to grow and harvest your own delicious (really!) spirulina in a modified aquarium. And we'll show you how to make a nifty self-contained electronic sound effects box called the Luna Mod.

One more thing: I'd like to remind you about Make: Projects, a terrific free service we set up for anyone to write and publish how-to projects. We recently ran a Karts and Wheels contest — you'll find winner Jeremy Ashinghurst's "Weekend Warrior" gravity racer on page 60. Check makeprojects.com for the upcoming robot competition!

Mark Frauenfelder is editor-in-chief of MAKE.

Arduinophytes and Maker Services

The Arduino is a very impressive and capable platform (MAKE Volume 25, "Getting Started with Microcontrollers"), but to imply that its programming language is easy for beginners is misleading. No beginner is going to be comfortable with a C-based development environment. The BASIC Stamp 2 or its derivatives are the best starting point.

For more advanced users, the Arduino is a logical choice. However, if you really want power and flexibility, the Propeller can't be beat. Its native language, Spin, is no more difficult than the Arduino's, and it has comparable hardware development platforms.

—Mark "DixieGeek" Andrews, Cullman, Ala.

AUTHOR TOM IGOE RESPONDS: In my article I admit to an Arduino bias. That bias is born of comparative experience. Having taught for many years with the BASIC Stamp, the BX-24, the PIC using PicBASIC, and Arduino, I disagree.

I've taught hundreds of beginners with Arduino, and they're comfortable with it. I've had similar reports from other teachers of nontechnical students. One of the keys is to keep the language clear and avoid the more complicated structures of C. For example, we've mostly been able to avoid pointers, which helps beginners a great deal.

You asked us to dive into the Arduino revolution (Volume 25). I wasn't sure what it was, but I bought the Getting Started with Arduino kit from the Maker Shed. As a 54-year-old woman, I'm probably not in your primary demographic, but I have to tell you I'm having a blast. I now have several books, lots of parts, and loads of creative enthusiasm. Thanks.

—Becca Begley, Louisville, Ky.

I'm a maker, age 12, and I really appreciate MAKE and makezine.com. They are both very helpful and informative, and I'm amazed at all the work that must go into developing them.

My coolest project, so far, is building my own microcontroller. I began with an ATmega Lite Development Kit, then constructed the parallel programmer, loaded the bootloader into the ATmega, and programmed it with the "Blink" program. One of my best sources for data (ATmega pinout, schematics, etc.) was MAKE Volume 25. This is the best magazine I have ever seen. Thank you so much.

—Chandler Watson, Hillsboro, Ore.

Visit MAKE's Arduino site at makezine.com/arduino for more great resources, including Phil Torrone's essay "Why the Arduino Won and Why It's Here to Stay."

I'm a MAKE subscriber who also subscribed to CRAFT magazine and was very disappointed when it ceased publication. I was mollified by MAKE's promise to publish more CRAFT-style articles (fiber arts, jewelry making, etc.), and I'm writing to express my wish that this promise not be forgotten. Part of fulfilling that promise could be a stronger emphasis on female makers and writers, to affirm their presence in the creative community.

Don't misunderstand me, you have a great magazine, but many readers came to it out of CRAFT, and I'd hate to see this element be lost.

—Adrian Miller, Rockaway Beach, Queens, N.Y.

MANAGING EDITOR KEITH HAMMOND RESPONDS: Thanks for sticking with MAKE, Adrian. We agree, and we're committed to bringing you a full range of makers and projects, from high tech to food, home, arts, and crafts.

In the 1970s I did a lot of home rocketry, including a rocket very similar to your $5 Heli-Rocket project (Volume 25). My model has a simpler release device, which might help people trying their own build; it's described at makezine.com/go/reedrocket. Great article!

—Reed Ghazala, Cincinnati, Ohio

I made my first project this weekend, the Stroboscope (Volume 24). Here are some photos I made with it: makezine.com/go/jamaal. I really look forward to more projects.

—Jamaal Montasser, Toronto, Ontario

I finally finished the Squelette amplifier (Volume 23, "Mini Chip Amp"). The instructions were easy to follow. I built my chassis from brass sheet and some cherry for the endpieces. The metalworking is poor, but hey, I'm a woodworker! Pictures of my amp are at brianandonian.com/LM1875chipamp.aspx.

The amp sounds really good: no hum, noise, or buzz. The headphone jack is a must. Ross, thanks for the plans. This was a fun build!

—Brian Andonian, Plymouth, Mich.

AUTHOR ROSS HERSHBERGER RESPONDS: Thanks for the pictures to drool over, Brian. I love the wood/brass combo. As for the sound, don't be shy about throwing it at some high-end speakers. More than one audiophile has been amazed by what these little chips can do.

Tell me it isn't true! I saw the latest issue of MAKE at a bookstore, and it was the same size as every other magazine. I thought *Popular Science* made a big mistake when they "upsized." Now you appear to be doing the same thing. I'm very sad. Nostalgic, too. But I'll continue reading MAKE.

—Roger Garrett, Honolulu, Hawaii

EDITOR-IN-CHIEF MARK FRAUENFELDER RESPONDS: The issue you're describing is actually our MAKE Ultimate Workshop and Tool Guide (makezine.com/2011/workshop). It's not MAKE magazine but an annual special edition.

Have no fear — we love the size of MAKE as much as you do and have no intention of changing the size!

I'm from a long line of makers, but my training is in urban public health. In Compton, Calif., most of the community's health problems (poor food, housing, violence) can be traced to the loss of factories and living-wage jobs for blue-collar workers. Talk all you want about retraining and scholarships, but we have a lot of good people with good skills who are makers, born to work with their hands.

We need a reboot, and I think the Maker movement is where our renaissance can come from. There are already communal shops being built, and Maker Faire and the internet are bringing marketplaces to more customers.

Why not take a page from big corporations and pass legislation to expand community shops? Provide "Maker Services" — a roving team from the Small Business Administration to help inventors micro-market, finance, and even patent (or Creative-Commons) their inventions. Create a special tax bracket for small-owner fabrication shops.

We need something to help creative minds and hands get their ideas to market, and create jobs while they're at it.

—Susan J. Hale, MPH, Santa Monica, Calif.

MAKE AMENDS

In Volume 25's *Secret Knock Gumball Machine*, page 95, we listed the wrong clear plastic globe from 1000bulbs.com. The correct part number is #3202-08020.

In Volume 25's *Sous Vide Immersion Cooker*, page 107, we listed the wrong clear acrylic box from Amazon. The correct part number is #B000NE9VJE.

In Volume 25's *Primer: Make and Use an Arduino*, page 65, we erroneously reversed power and ground for pins J-7 and J-9. The text should read: "To connect the chip to power and ground, jumper [...] J-7 to right ground, and J-9 to right power (Figure G)." Figure G is correct.

In the calculation in Volume 25's *Remaking History*, the crank diameter is 6", not the "crank handle" (radius). The equation is correct for a 6" diameter and 3" handle.

Jan	Feb	Mar
Apr	May	Jun
July	Aug	Sept
Oct	Nov	Dec

MAKER'S CALENDAR

COMPILED BY WILLIAM GURSTELLE

Our favorite events from around the world.

Maker Faire Bay Area 2011

May 21–22, San Mateo, Calif.

There's always something new and terrific at Maker Faire Bay Area, the world's largest DIY festival, sponsored and organized by MAKE magazine. The two-day, family-friendly event gives everyone a chance to create, learn, invent, craft, recycle, think, play, and be inspired. makerfaire.com

MAY

» Science Rendezvous

May 7, cities across Canada

More than 1,500 scientists and technologists, representing the cream of Canada's scientific community, throw open the doors of their laboratories to talk "aboot" their work. This go-round celebrates 2011 as the U.N.-designated International Year of Chemistry. sciencerendezvous.ca

» Friday Harbor Labs Open House

May 14, San Juan Island, Wash.

Here's an opportunity to meet the scientists and students of a renowned marine biology field station and check out their research and teaching facilities. The FHLOH showcases marine science research, including demonstrations and Q&A. There will be marine plants and animals, microscopes, research ship tours, and activities for visitors of all ages. depts.washington.edu/fhl/events.html

» NYU Music Technology Open House

May 14, New York, N.Y.

Arduino-lovers and electronic music fans rejoice — it's a night of music, sound, images, and thought brought to you by NYU's music technology community. Check out cutting-edge gadgetry as NYU's creative students present performances and interactive demonstrations. makezine.com/go/nyu

JUNE

» Maker Faire North Carolina

June 18, Raleigh, N.C.

From James Bond-worthy electronic gizmos to Martha Stewart-quality "slow made" foods and homemade clothes, Maker Faire North Carolina celebrates things people create themselves. Maker Faire NC is an up-and-coming, fully sanctioned event, planned and coordinated by Raleigh/Durham makers. makerfairenc.com

» National Threshers Association Reunion

June 23–26, Wauseon, Ohio

Every maker loves a parade, especially when it's made up of lovingly restored, antique steam- and gasoline-powered vehicles. The 67th annual reunion features approximately 50 steam engines, in addition to more than 100 gas tractors and gas engines. Daily demonstrations include threshing, sawmills, shingle mills, plowing, and a machinery parade. nationalthreshers.com

» Eyeo Festival

June 27–29, Minneapolis, Minn.

This new three-day conference includes a wide variety of talks, demos, labs, and workshops. Subjects vary widely and touch on all topics digital, including digital art and music, coding, Arduino, 3D printing, and more. Sessions include a number of presenters who've been featured in MAKE magazine and on *Make:* Television. eyeofestival.com

JULY

» The International Bognor Birdman

July 16–17, Bognor Regis, England

Flap your way over to England's International Bognor Birdman and watch in mock horror as stalwart human "birdmen" attempt to fly their outlandish homemade contraptions off the end of a pier in front of thousands of people — for glory and perhaps a bit of prize money. birdman.org.uk

» Maker Faire Detroit

July 30–31, Dearborn, Mich.

Could there be a more fitting place for a Maker Faire than the Motor City? No way! Last year there were more than 325 sponsors, exhibitors, and performers in attendance. This year should be even bigger and better. Don't miss it. makerfaire.com/detroit/2011

✱ IMPORTANT: Times, dates, locations, and events are subject to change. Verify all information before making plans to attend.

MORE MAKER EVENTS:

*Visit **makezine.com/events** to find classes, exhibitions, fairs, and more. Log in to add your events, or email them to **events@makezine.com**. Attended one of these events? Talk about it at **forums.makezine.com**.*

Projects in Your Mailbox

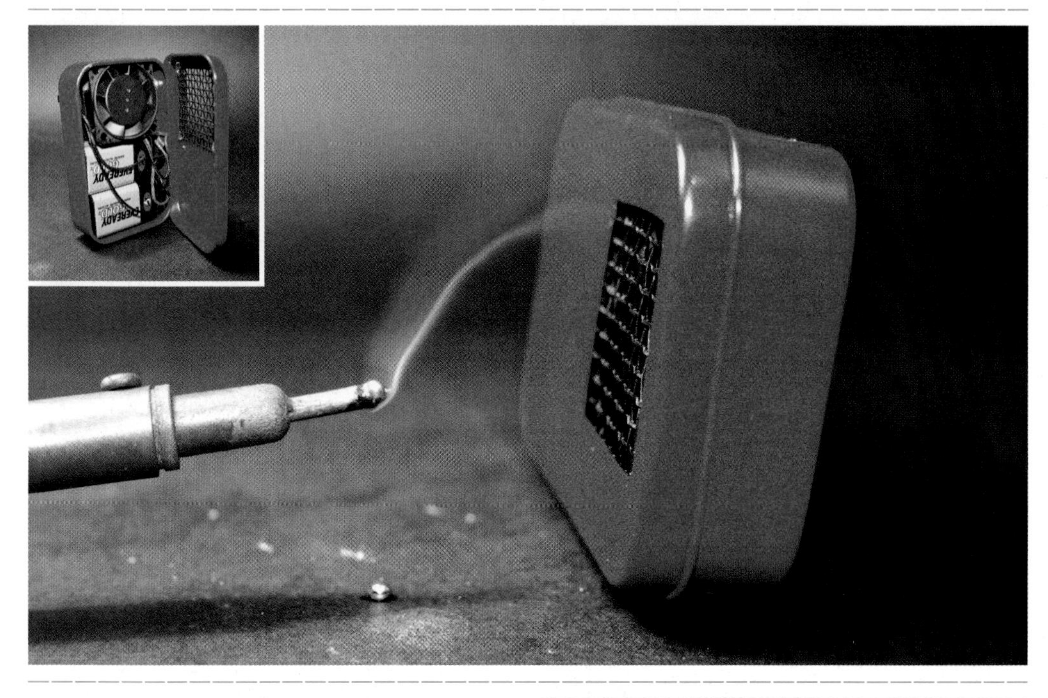

WHILE SETTING UP FOR OUR FIRST Maker Faire, I went on a supply run to Jameco Electronics, a short drive from the fairgrounds near San Francisco Bay. As I pulled into the parking lot, I saw four grinning fellows sporting "Make: Void your warranty" T-shirts piling out of a car and kneeling in front of the familiar Jameco sign for an impromptu group photo.

These makers — and thousands like them — didn't need to be told about the fit between MAKE and Jameco. They just got it.

So when MAKE sought to make it easy for readers to buy the supplies they needed to tackle projects found in the magazine and at Make: Projects (makeprojects.com), our first stop was Jameco. Like that carload of exuberant makers, the team at Jameco got it.

We're proud to announce that Make: Projects readers can now conveniently purchase just about everything they need for their next project directly from the Make: Projects website. Select the complete project bundle, or just the components you need, then click on the shopping cart link to complete your order in a single, easy Maker Shed transaction.

A perfect example is Marc de Vinck's Mini Fume Extractor project from MAKE Volume 19, released on Make: Projects at makezine.com/go/fume. We're adding hundreds more projects in the weeks to come. And when you drop by the Maker Shed (makershed.com) you'll find thousands more Jameco products identified with their familiar icon.

Mini Fume Extractor Parts Bundle

$39.95 Product code JMBUN07 makershed.com/fume
NEW: Just click on the cart in Make: Projects (makezine.com/go/fume) and your Mini Fume Extractor parts will be speeding toward your mailbox.

Dan Woods is MAKE's associate publisher and general manager of e-commerce.

Photography by Marc de Vinck

Made On Earth

Reports from the world of backyard technology

Ultimate Hot Wheels Track

Whether you love or hate him, California artist **Chris Burden** is a genius — he makes a living playing with toys, on an epic scale.

Burden's kinetic sculpture *Metropolis II* is a mesmerizing cityscape where 1,100 toy cars blaze down 18 lanes of freeways in endless loops. The work took Burden, his chief engineer **Zak Cook**, and ten assistants four years to build in his Topanga Canyon studio.

Los Angeles County Museum of Art director Michael Govan promptly dubbed it "a portrait of L.A." and secured it for display this fall.

A forerunner sculpture used Hot Wheels cars and tracks, but these proved unreliable at high speeds, so Burden switched to custom cars and plexiglass roadways. Industrial conveyor belts grab the cars magnetically, hoist them high, then release them to careen through an architecture of Unistrut framing, Lego, Haba Blocks, and Lincoln Logs.

Burden, 65, made his name in the 1970s with risky performance art (he had himself shot with a rifle and crucified on a Volkswagen, among other stunts) before turning to installations and sculptures exploring science, technology, and politics.

Metropolis II has layers of meaning — there's the hurry-up-and-stop traffic going nowhere, the ceaseless racket of 100,000 cars passing every hour, the portent of robotic vehicles. "This idea that a car runs free, with a driver in control who can accelerate at will, is soon going to be a notion of the past," Burden predicts. "All cars will have a digital slot. You'll be able to go 200 or 300mph on the freeway."

It's also the latest of Burden's great big toys. He's built a 65-foot skyscraper from Erector parts, and launched a self-navigating sailboat. Next he's building a two-story bridge of scaled-up Anker Stone toy blocks.

—Keith Hammond

18 Lanes: makezine.com/go/burden

In the Lion's Den

Close-up photography of African wildlife has always been an extreme sport. And capturing a unique photo is even more challenging. But then along came a tiny robot called BeetleCam.

Built by U.K. photography duo **Will** and **Matt Burrard-Lucas**, the BeetleCam is a modded 4WD robotic buggy manufactured by Lynxmotion and mounted with a DSLR camera. The brothers ordered their buggy with off-road tires to help it traverse Africa's uneven terrain. Then they stuffed it with two 7.2-volt, 2,800mAh NiMH battery packs that provide a daylong operating time — a must since wildlife photography involves a lot of waiting, followed by much more waiting.

With some creative hacking, the camera is operated by a free channel on the Hitec 6-channel radio controller. A relay switch converts the signal from the R/C receiver to one that triggers the camera's shutter release cord. The camera, originally a Canon EOS 400D, controls two flash modules using a split off-camera flash cord.

The robot cost $500 and took just a month to prototype. After they stabilized the camera and camouflaged the bot, it was ready for a trip to Tanzania — where it was promptly mauled by a lion and carried off into the bush.

While the camera was destroyed, the memory card and robot chassis were intact. And the downloaded photos proved BeetleCam a success. The unit was quickly upgraded with a Canon EOS-1D Mark III and sent on its way. (Lions were to be avoided.) Oddly, another of Africa's most dangerous animals — the African buffalo — proved to be not only cooperative but curious about the tiny critter.

The team is now working on version 2.0 of BeetleCam. Will says they also "intend to build a lion-proof version." —*Jerry James Stone*

BeetleCam Bros.: burrard-lucas.com

Photograph by William and Matthew Burrard-Lucas

RFID Radios

Matt Brown's laser-cut RFID (radio frequency identification) radios were first conceived while he studied interaction design at the Umeå Institute of Design in Sweden. A design firm he was applying to at the time asked him to do a "personal project around music."

He approached the challenge excitedly, thinking about ways he might re-construe classic forms with new technologies. And although he ultimately wasn't offered the job, an innovative combination of RFID, internet radio, and laser-cut technologies was born.

Brown's idea is to affix an RFID chip inside a laser-cut, flat-pack paper radio, and then pair the radio with a speaker base with an RFID reader.

Each radio would be designed by a different musician or artist. When the radio is placed over the speaker, the station shifts to that artist's pre-selection.

"This system tries to add a little bit of fun to internet radio and give people a connection with the artists they choose," says Brown.

Inspired by things that are neglected or obsolete, Brown strives to find ways to make them useful and enjoyable again. Like many of us, he feels that certain vintage aesthetics are superior to much of what's produced today.

"It just seems like people got away with more interesting designs in the 50s, 60s, and 70s," Brown says. "I go to a lot of thrift stores and spend a lot of time on eBay. You start to get this library of details in your brain, and then when you sketch out a new design it's a strange combination of all these details."

The RFID radio idea was Brown's first experimentation with a laser cutter and paper, and it seems to have lent itself well to his minimalist, retro designs.

—Thomas Wilson

» **More of Brown's Designs:** skrov.com

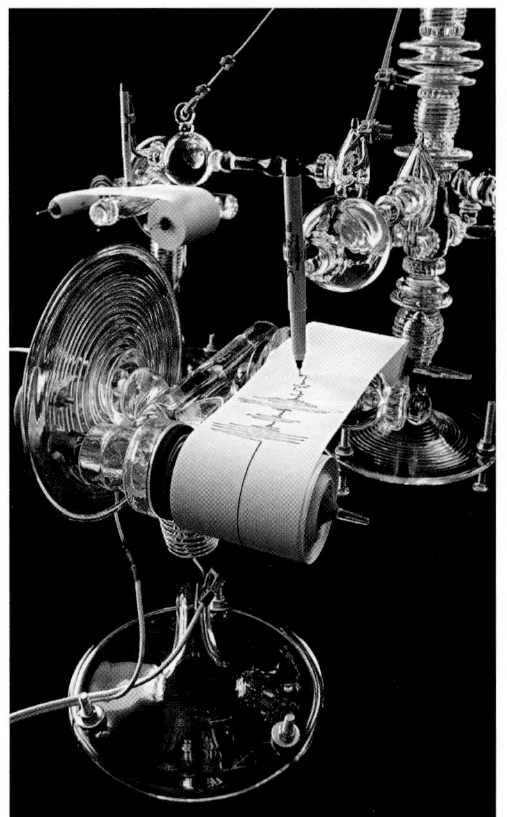

Delicate Machinations

Andy Paiko is a glass blower known for his intricate vessels that deftly mix nostalgia and modernity, and look equally at home in the Mütter Museum of medicine or the Guggenheim. Yet he strives to create more than just beautiful objects. A lover of functionality, he is passionately reinventing how glass behaves and how it's ultimately perceived.

Paiko has painstakingly re-created intricate antique machines almost exclusively out of glass, including a seismograph, a spinning wheel, weight scales, and even a large-scale Ben Franklin-esque glass armonica — and yes, they all work.

It's impossible for photos to capture the magic of these machines, but a brief video from Oregon Art Beat shows a few of them in motion. It's mesmerizing to watch his glass gadgets perform real-world tasks while hearing them tinkling and clinking as they move.

Paiko's breakthrough moment came years ago when a college professor, upon looking at his array of staggeringly beautiful pieces, asked him, "Have you asked yourself whether or not the world really needs another glass vase?"

From that moment on, Paiko strove to create glass that shattered expectations.

Though his favorite artists include Marcel Duchamp, Max Ernst, Tim Hawkinson, Tom Friedman, and recently Martin Puryear, Paiko is most influenced by the real world.

His scientific heart is warmed by antiquated technology and uniquely engineered machines. He's also inspired by mycology, botany, humanism, and, he says, "anyone who makes or does something original in their life without having to try too hard."

—Stacey Ransom

» **Paiko Glass:** andypaikoglass.com

In Motion: makezine.com/go/artbeat

Pooled Resources

When **Dennis** and **Danielle McClung** purchased a foreclosed home in Mesa, Ariz., they found themselves with an empty, run-down swimming pool. But rather than sink money into filling it in or restoring it, they transformed the backyard hazard into a desert greenhouse.

Their 480-square-foot, 15-foot-tall Garden Pool (GP) is the first step in the family's quest toward self-sustainability. After covering the in-ground concrete pool with an anti-UV treated tarp draped over flexible PVC rods, they created an experimental farm that utilizes solar power, permaculture, organic farming, hydroponics, aquaponics, and biofiltration.

At the pool's deep end is a 6,703-gallon pond teeming with edible tilapia fish. Pumps supply unfiltered pond water to the GP's fruits and vegetables — blackberries, eggplant, broccoli, and spinach — grown in buckets, rain gutters, kiddie pools, and on trellises.

Filtered by the garden plants, the water then trickles back to the fishpond. The GP's chickens lay fresh eggs daily, and their waste fertilizes the garden plants and the algae that grows on the bottom of the pond as fish food.

The McClungs' invention generates enough food to feed their family of four, and the entire natural process pretty much cares for itself.

"I go out in the morning and make sure the chickens are fine," says Dennis. "I check on the flow of the pumps and see how the solar battery bank is doing. I don't have to weed, water, or compost."

To assist budding pool gardeners, Dennis, a web designer, documents the GP online and offers tours. Despite its unusual appearance (picture a makeshift bio-research lab), the neighbors seem to love it. "Some have come over during our tours and are just blown away," he says. "They think it's great."

—Laura Kiniry

More GP Photos: gardenpool.org

Orbital Illusion

Fairgoers at the 2010 Bay Area Maker Faire were mesmerized by a massive, steel-and-glass orrery gently rotating in the main hall. At the controls was artist **Orion Fredericks**.

Like much of his work, *Fata Morgana* (a term for a complex mirage) was inspired by a dream. Fredericks, 34, sought to create a kinetic sculpture that filled the viewer with the same sense of awe and wonder he'd experienced.

First he crafted a small but more complicated version and learned from its mistakes. Then he spent a month drawing and planning, another month gathering materials, and a final month of solid fabrication. What he ended up with is a fascinating play on light and illusion, a mechanical representation of a solar system — not ours but the one he envisioned.

At the core of the sculpture are a 1,000-watt full-spectrum light bulb and a 24-volt wheelchair motor that drives the primary shaft. Power is transferred through gears and pulleys to four friction drive wheels that rotate two tiers of the sculpture clockwise and two counterclockwise. Passively kinetic elements are carefully balanced throughout.

A second motor controls the optical beam splitter. When the central light reaches the outermost 3-foot-diameter rotating lenses, *Fata Morgana* projects an illusion of itself around its perimeter.

Materials were donated or scavenged in the "nooks and crannies" of Oakland, Calif., Fredericks says. And though it's hard to imagine, since it measures 15 feet tall and 24 feet wide, *Fata Morgana* is actually a prototype for a larger, more complicated piece to come.

Like a solar system, the sculpture has a gravitational pull of its own. The artist notes, "People usually want to sit under or close to it."

—*Goli Mohammadi*

» **Orion Fredericks:** orionfredericks.net

***Fata Morgana*:** vimeo.com/12452045

Photography by Branca Nitzsche

Ice Music

"I go up to a very high place in the mountains. In the shadows next to a stream I set up my little workshop under a couple of tarps," says, Taos, N.M. ice-instrument maker **Tim Linhart**.

While Linhart creates his fully functioning 14-piece ice orchestra, his wife, **Birgitta Linhart**, oversees construction of the Celestial Sphere concert hall — a complex of giant igloos with holes at the top for audience and performers' body heat to escape.

"We make the orchestra, concert hall, install laser lighting and the sound system, and are putting on music six weeks after we start ... it's a super explosive time frame," Tim explains.

Unlike other makers who carve instruments from blocks of ice, Linhart uses an additive method to create many of his. (His giant xylophones and a few structural components are exceptions.) He packs a slush of snow and water onto various forms in the shape of drums, pipes, and his ingenious Rolandophone (a unique "compression instrument" that's a sort of hybrid marimba, drum, and pipe organ). When the slush hardens, it becomes a "white ice" shell, which is then separated from the form.

Linhart's hand-sculpted stringed instruments have standard necks and a strip of wood running down their bodies. They are delicate; the ice is shaved very thin so that it's vulnerable and flexible under the tension of the strings. "It's like playing chicken with an explosion: the closer you get to beautiful sound, the closer you are to the explosion."

Other makers let their ice instruments melt away, but Linhart takes another approach. "All the players can hammer their favorite instrument to bits. ... There's a joy in destroying perfection. It's only ice — it'll come back next year." —*Kim Bailey*

» **Celestial Concert:** icemusic.us

Live from MAKE!

THIS YEAR WE NOT ONLY UPGRADED OUR WEBSITE AND MOVED "INTO THE CLOUD" — we hope you're enjoying the improved performance, social media, and commenting tools — we also launched these great new programs online. Check them out!

☼ **Make: Live** In partnership with Digi-Key, we're streaming show-and-tell video with our *Make: Live* program. Twice a month, hosts Becky Stern and Matt Richardson explore projects from MAKE and the wider maker world, presented in-studio and via Skype by the makers. Check out episodes on Arduino, soldering, bikes, and more. makezine.com/live

☼ **PT's Got Issues** We sent Editor-at-Large Phillip Torrone on a grand adventure spelunking through his own multitasking brain, and he's causing a ruckus. PT is an icon in the open source/hardware hacking community, and he's one smart cookie who does a lot of cogitating about where the DIY world is heading. Don't miss "Why the Arduino Won" and "Sony's War on Makers." makezine.com/go/pt

☼ **Treasure Hunt** Make: Online has six years of great original content in our vaults — but who wants to rummage through vaults? Two gems not to miss:

» Make It Anywhere with a Mobile Lab Steve Roberts' full workshop in a trailer. makezine.com/go/mobilelab

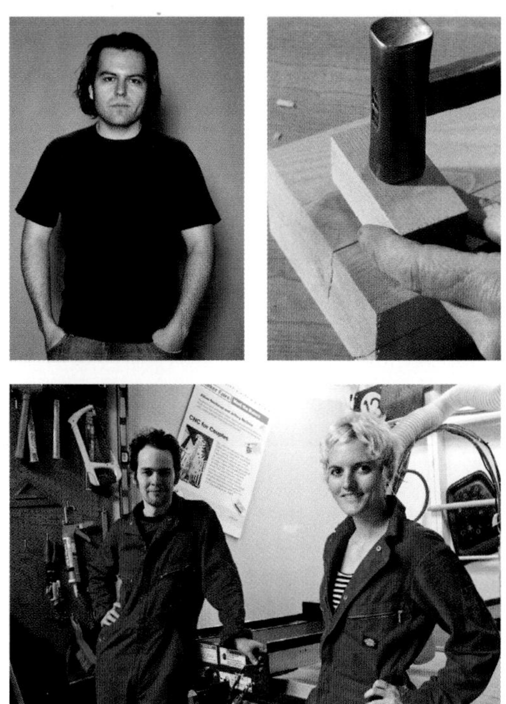

» Maker Pro Turning maker passions into livelihoods. makezine.com/go/makerpro

☼ **Make: Skill Sets** What every maker needs to know in electronics, woodworking, machining, and more. Gain core competency in 12 basic DIY skill sets! Just check the "right rail" each month at blog.makezine.com.

☼ **Build a CoasterBot** Get started in botbuilding, using project materials from our popular 2010 contest. Jameco still offers the parts bundle! makezine.com/robotbuild

---

Gareth Branwyn is editor-in-chief of Make: Online.

Photograph by Jason Forman (Phil Torrone)

DIT: Raising Our Collective Barn

I HAVE A FRIEND WHO STARTED USING the term DIT ("do it together") instead of DIY, and I love the notion. It's a rare project that's completely DIY, and certainly the projects that are the most fun involve collaboration — even if it's just with the cranky old guy at the hardware store who ties your whole project together with some piece of ancient wisdom that should have been obvious but wasn't.

I've been building an attic playspace for my toddler recently. What I really mean is *we've* been building it. At least a dozen friends and family have contributed, from framing the space to painting murals to building a ladder.

When you build it together, your learning is amplified — multiplied by the number of hands.

What's beautiful is how much everyone has enjoyed being part of the build. Sure, they get paid in tea, cookies, and cold beer (the currencies of DIT), but they also look forward to sharing the space with us in the future. When unexpected visitors drop by, I enlist them and they accept with gratitude. It is indeed a nice day to whitewash the fence.

It's like barn raising. Hundreds of years ago, this is the way many things were built — by depending on each other, investing in our community through shouldering a fair load.

My own culture (Australian) still idealizes the notion of "mateship": sticking together through thick and thin, leaning on one another when there's hard "yakka" (work) to do.

It's astonishing how much easier hard yakka is when you're doing it with friends. My wife has about 50 cousins (no, really), and at family events there's never a shortage of labor. This swarm of young people seems to compete to see who can get the most work done in the shortest time. That's the beauty of barn raising — our social drive to display our prowess is engaged, and stuff gets done quickly with smiles and laughs.

Makers are starting their own groups, collectives, and makerspaces all over the world. We're feeling the high of DIT. Everyone can contribute time, skills, or even just muscle, knowing it will be paid back at the end of the metaphorical day.

An ancillary benefit is learning. When you make something alone, you're learning by yourself, through books or a search engine. When you build it together, your learning is amplified. Subtle things are absorbed: a better way to swing a hammer, to finish a routing job, to solder tiny surface-mount components. Everyone on the job site has a unique set of skills, and the amount to be learned is multiplied by the number of hands.

There's a lot that needs fixing and a lot that needs building. In cities, I see so much unused and unloved space: abandoned lots, rotting industrial buildings. I hope we find ways to reclaim these spaces. I suspect they wouldn't take much to revamp into gardens and community centers, places we want to be. They just need some barn-raising love.

That may be hippie heresy, but hey — the social, economic, and political rules are changing. Why don't we exploit the moment and build the future we want together? We have the best tools in history for social organization (thanks, Twitter; thanks, Facebook). Let's apply them to the places around us. Let's invest collectively in our infrastructure. Do your friends a favor — raise a barn, and involve them.

Saul Griffith is a new father and entrepreneur. otherlab.com

BY FORREST M. MIMS III

Ultra-Simple Sunshine Recorders

SUNSHINE IS ESSENTIAL FOR PLANT growth, which is one reason scientists have developed so many methods to measure it.

In a future column I'll show how to measure sunshine electronically. But first, let's enter the MAKE time machine and zip back to 1838, when instrument maker T.B. Jordan made the first known automatic sunshine recorder.

Jordan wrapped a strip of silver chloride photographic paper around a clock-driven cylinder, and mounted it behind the mercury column of a barometer. The photo paper was exposed to sunlight as it rotated past an aperture, and this provided a record of when clouds blocked the sun. The width of the exposed portion of the paper was controlled by the height of the mercury. Thus the instrument was both an early barometric pressure recorder and the first sunshine recorder.

In 1840, Jordan designed a second sunshine recorder whose photographic cylinder was exposed to sunlight passing through a pinhole in a second cylinder that rotated about the photo cylinder once in 24 hours.

You can make simple versions of Jordan's sunshine recorders using cyanotype paper. It's easy to use, and its image is preserved by simply immersing the paper in water.

The Cyanotype (Blueprint) Process

When he wasn't flying aircraft for the U.S. Air Force, my father was a civil engineer and an architect. Detailed architectural drawings, imaged on cyanotype paper and known as blueprints, were displayed on his office walls.

Today, *blueprint* means a detailed plan or a physical model of something, but originally it meant a cyanotype print of an architectural or engineering illustration. Blueprints were created by first making an original drawing on a sheet of translucent paper. This drawing was placed over a sheet of blueprint paper and exposed to ultraviolet light. The blueprint paper was then developed in water rendered slightly acidic. The paper exposed to the UV light became a rich blue, while the paper shadowed by the drawings became white.

Easily record sunshine and cloud cover by using cyanotype blueprint paper with a film canister or a cheap quartz clock.

The blueprint process was discovered by Sir John Herschel in 1842 and has remained essentially unchanged ever since. Herschel described his discovery in his paper "On the Action of the Rays of the Solar Spectrum on Vegetable Colors, and on Some New Photographic Processes," published in *Philosophical Transactions of the Royal Society* (June 1842). You can make your own blueprint paper, or buy it from hobby and craft stores. I use SunArt paper by TEDCO (tedcotoys.com). It's available in packages of 15 sheets measuring 4"×6", 5"×7", or 8"×10".

Ultra-Simple Pinhole Camera Sunshine Recorder

The simplest sunshine recorder is a light-tight enclosure in which blueprint paper is placed opposite a pinhole. This fits the definition of a true camera.

Before sunrise, the recorder is mounted in a fixed position so that the pinhole aims at a point about halfway between solar noon (where the sun will be at its highest point in the sky) and the southern horizon (northern horizon if you're in the Southern Hemisphere). After sunset, the recorder is

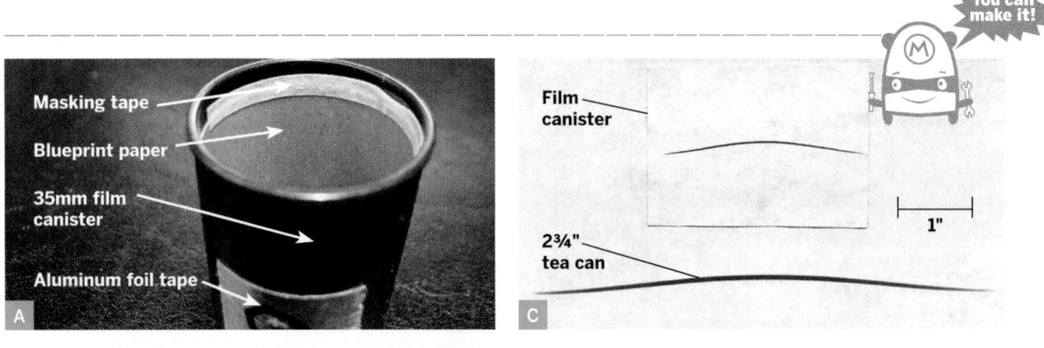

Fig. A: Pinhole sunshine recorder made from a plastic 35mm film canister. The blueprint paper is held in place by a thin strip of masking tape.

Fig. B: The film canister sunshine recorder can be mounted on a tripod by boring a ¼" hole in the lid and securing it to the tripod screw with a 4-40 nut.

Fig. C: Sunshine traces made on the same cloud-free day by a 35mm film canister recorder with a 0.5mm pinhole and a tea can recorder with a 1mm pinhole.

opened and the blueprint paper is washed in water mixed with a few drops of lemon juice to enhance the color. The sun's track across the sky is preserved on the paper as a deep blue arc. Any interruptions in the arc indicate when clouds passed in front of the sun.

You can easily transform a plastic 35mm film canister into a sunshine recorder. Bore a ¼" hole in the side, then place a small square of aluminum foil tape over the hole, and use a needle to make a pinhole through the foil. Roll a 1¾"×2⅜" piece of blueprint paper into a cylinder (sensitive surface facing inward) with a gap somewhat wider than the pinhole, then insert it into the canister so the pinhole is centered between the edges of the paper.

For best results, hold the paper in place with a thin strip of masking tape applied to the upper edge. Be sure to install the blueprint paper indoors to avoid exposing it to UV from sunlight. Then snap on the lid.

Your pinhole sunshine recorder is now ready for use. Simply tape it to a fixed support so that the pinhole is pointed as described previously. For even better results, mount it on a camera tripod. This is easy; bore a ¼" hole through the canister's lid, then attach the lid to the threaded bolt of a tripod with a 4-40 nut. Then load the canister with blueprint paper and press it down into the lid.

If a pinhole recorder is left out in the rain, moisture can get inside. This shouldn't affect the previously formed sunshine trace, but it can cause the blueprint paper to wrinkle or the image to be prematurely preserved.

Clock-Driven Sunshine Recorder

Quartz-controlled analog clocks are available for as little as $10. You can easily make a sunshine recorder from such a clock. While this recorder uses a very small sunlight aperture, it's not a true imaging pinhole camera, since the aperture is in motion as it scans a speck of sunlight across a sheet of blueprint paper.

Start by removing the clock's housing; then remove the clock hands by carefully pulling them straight up from the face of the clock.

Cut a disk of black paper slightly smaller

Photography by Forrest M. Mims III

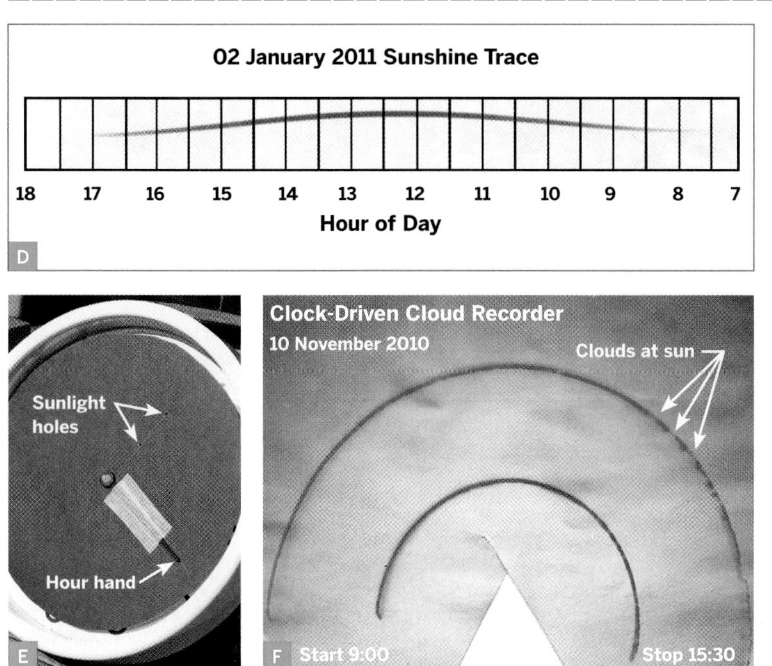

Fig. D: Digitized sunshine trace from tea can recorder in Figure C annotated with 30-minute grid lines. The center grid line was aligned with solar noon (12:36 p.m. on Jan. 2, 2011). Grid lines were equally spaced from noon to sunrise and noon to sunset. Daily sunshine times can be found at sunrisesunset.com.

Fig. E: Quartz analog clock-driven recorder with two 1mm sunlight apertures. The black paper disk was taped to the clock's hour hand. Blueprint paper was taped to the clock face behind the black paper disk.

Fig. F: Dual sunshine traces made with the clock-driven sunlight recorder. Note the interruptions in the traces caused by clouds blocking the sun.

than the clock face, and make a ¼" hole at its center. Place the axis of the hour hand over the hole, and tape the hand to the disk. Make a 1⁄16" hole in the disk 2" from the center. This aperture will allow sunlight to strike the blueprint paper as the black disk rotates over it.

Cut a rectangular sheet of blueprint paper to cover the clock face's top half. Again, do this indoors so that no sunlight or UV light strikes the paper. My clock face is 7¾" in diameter, so I cut a 3½"×5" piece of SunArt paper, with a V-shaped notch in the bottom to make room for the clock drive axis.

Place the blueprint paper over the clock face and secure it with masking tape. Finally, place the black disk over the clock face with its taped hour hand facing up, and its sunlight aperture aligned at a morning hour, and press the hour hand onto the clock axis. My 3½"×5" sheet allowed six hours of sunlight recording, so I set the aperture to the 9 a.m. setting.

Place the clock outside so it faces the sky as described earlier. Retrieve it at sunset and remove the black disk, then remove the blueprint paper and preserve its image by immersing it in water with a few drops of lemon juice.

Going Further

To make your own cyanotype paper, see the instructions at makezine.com/go/cyanotype. In Google Books, you can even find recipes and methods in Herschel's original paper.

Is the density of the blue color of a sunlight cyanotype trace proportional to the intensity of the sun's ultraviolet light? Finding out could be an excellent science project. Digitize a sunlight trace, then use photo software to determine the blue density at uniform intervals along the trace. Compare these data with the UV for your site measured by a Solarmeter (solarmeter.com) or similar UV radiometer, or by the nearest radiometer in the U.S. Department of Agriculture's UVB monitoring program (makezine.com/go/usda_uvb).

Finally, pinhole cameras can be used with ordinary photo paper to make spectacular sunshine traces of up to six months' duration. See Tarja Trygg's how-to instructions at www.solargraphy.com and Justin Quinnell's images at pinholephotography.org.

Forrest M. Mims III (forrestmims.org), an amateur scientist and Rolex Award winner, was named by Discover magazine as one of the "50 Best Brains in Science." His books have sold more than 7 million copies.

Moral Suasion

THE LIFE OF A MODERN-DAY MAKER is greatly eased by the abundance of free or cheap services for hosting communities, files, communications, and computation. But for all the promise of cloud computing, there's plenty of peril, too, especially for anyone doing anything disruptive.

Will your "cloud" evaporate the second your project starts attracting legal threats? Does a service provider with a million customers care about your customs enough to keep you online even if it means risking a police raid, subpoena, or denial-of-service attack?

The ongoing WikiLeaks fight is a wake-up call for anyone who's been blithely relying on the cloud. It only took a few days for WikiLeaks to become a digital refugee, slogging from one service provider to the next, trying to find someone with enough backbone to keep it online in the face of legal threats, political intervention, and mysterious traffic-floods from persons or governments unknown.

But WikiLeaks wasn't without its defenders. "Hacktivists" operating under the Anonymous banner organized mass denial-of-service attacks on Amazon, PayPal, Visa, MasterCard, and other firms that kicked WikiLeaks out or refused to process their payments.

Distributed denial-of-service (DDoS) attacks are a strange beast, unique to the online world. Some DDoSes are the work of millions of users acting in concert to flood a server with so much traffic that it falls over. More commonly, DDoS attacks are the work of vandals or crooks who use clouds of hacked PCs to attack their targets.

Some hacktivists argue that their DDoS attacks are comparable to the civil-rights-era sit-ins — after all, a wall of activists blockading the doors to a "whites-only" lunch counter is a kind of denial-of-service attack.

I think they're wrong. I grew up in the antiwar movement and participated in my first sit-in when I was 12. Sit-ins are a sort of denial of service, but that's not why they work. What they do is convey the message: "I am willing to put myself in harm's way for my beliefs. I am willing to risk arrest and jail. This matters."

This may not be convincing for people who strongly disagree with you, but it makes an impression on people who haven't been paying attention. Discovering that your neighbors are willing to be harmed, arrested, imprisoned, or even killed for their beliefs is a striking thing.

A wake-up call for anyone who's blithely relying on the cloud.

And that's a crucial difference between a DDoS and a sit-in: participants in a sit-in expect to get arrested. Participants in a DDoS do everything they can to avoid getting caught.

If you want to draw a metaphor, DDoSers are like the animal rights activists who fill a lab's locks with super glue. This is effective at shutting down your opponent for a good while, but it's a lot less likely to draw sympathy from the public, who can dismiss it as vandalism.

One thing is clear: those of us who don't supply our own digital infrastructure depend on intermediaries who are increasingly willing to roll over at the slightest pressure.

It's time to start devoting some of our creative attention to ways of clearing away the choke-points and leaning back on those companies that are getting leaned upon by powerful, established forces.

Cory Doctorow's latest novel is *Makers* (Tor Books U.S., HarperVoyager U.K.). He lives in London and co-edits the website Boing Boing.

Maker
MANUFACTURE YOUR PROJECT

(And make a living doing what you love.)

By Mitch Altman

What would your life be like if you spent your time working on your own project, making enough money to live the life you want to live? Hold that image and think about what it would take to get there.

First of all, you have to pick an idea that you love. Remember, this project is going to take over your life!

When I came up with the notion of TV-B-Gone (makezine.com/go/tvbgone) universal remotes, I couldn't know how it would change my life, and a bit of the world. I learned a great deal from the process of bringing it from a mere idea to a successful reality. By sharing my experience, I hope I'll help make yours that much easier.

The steps I've outlined in this article may not seem all fun. But remember, each step along the way, the process is based on a project you totally love, and you know deep inside that you need to put it out into the world. The lows are something to get through to learn from.

The highs are incredible! You'll meet people with ideas to support your idea. You may travel places where you'll meet more incredibly cool people. You'll learn so much about yourself and the world as you play parent to your "child" and watch it find its own place out there, knowing that you can take these experiences and use them to make your idea (and your next ideas) work better, with fewer problems, and more fun. The process is incredibly fulfilling.

MAKE contributor Mitch Altman gives workshops teaching people to make cool things with microcontrollers, is CEO of Cornfield Electronics, and is co-founder of Noisebridge hackerspace. He is currently writing a book based on his workshops.

“There’s no guarantee that if you do what you love, you will make a lot, or even earn a living from it. But if you don’t, what is guaranteed is that your ideas will remain only dreams, and those dreams may fade.” *—Mitch Altman, inventor of the TV-B-Gone Universal Remote*

Photograph by Cody Pickens

1. BUILD A WORKING PROTOTYPE (OF PRODUCT AND PACKAGE).

You may not be an expert at everything. To build my prototype, I coordinated a bunch of friends to help with the tasks they're good at: to do a printed circuit board (PCB) layout, to design the case, and to create mechanical drawings for the remote.

I also tasked friends to create graphics for the packaging and website, to design, program, and host the website, to incorporate the company (a lawyer friend), to set up accounting software, to do publicity, and to do fulfillment and shipping. You're the one who will coordinate all these people's tasks into a unified vision.

In addition to these, some things needed to be done that were a surprise to me:

Barcode — If you want to sell your product in stores, you may need a barcode printed on the packaging. GS1 (gs1.org/barcodes) is the organization that keeps track of all companies' barcodes.

Package design — Nearly all products need to be sold in some kind of packaging, and there are many shapes and sizes to choose from. The one I chose is called a clamshell, since it's hinged and opens up from the bottom, somewhat like a clam.

I found out that the size of the package is important. If a package is too small, that may limit the price people are willing to pay for it. But if a package is too big, it costs much more to ship and to store.

Carton design — The factory where your product is manufactured needs to put the finished products into shipping cartons. How many units do you put in each carton? If you want to sell wholesale, you have to pick a number that buyers want. I chose 20.

Someone needed to design the carton so that it fit these 20 in such a way that they wouldn't get damaged in shipping. A designer also needs to size the carton so that an integral number will fit perfectly on a shipping pallet in all directions, and needs to know what markings to put on the side of the carton to please customs officials and tell recipients what's inside. Fortunately, my manufacturer was able to design my shipping carton for me.

2. OBTAIN FUNDING.

Manufacturing your project takes money, but not as much as you might think. To develop TV-B-Gone took about 18 months of my time, and about $2,000 of my dollars, including a trip to China to check out the manufacturer I finally chose.

There was an additional $10,000 in non-refundable engineering fees (called NRE fees) that I had to pay my manufacturer for all the tooling necessary to make the plastic case and the plastic packaging, and for setting up the cool artwork for the packaging, and such.

When it's time to truly press the Spend button and get your first units manufactured, you need to come up with the money to pay for each unit. Depending on your project, this can vary from a fraction of a dollar to several dollars apiece. Since the minimum quantity most manufacturers will consider for production is several thousand, even if your product only costs you a few dollars each, the cost multiplies quickly.

Where do you get the money? I borrowed from friends and family. The advantage of this approach is that everyone who gave me money likes me, likes my project, knows I'll do my best to pay them back, and, luckily for me, did not require a large amount of interest in return.

Another route is "angel" investors who have spare cash to invest in cool projects. If you need millions, I'd suggest rethinking the cost of your project, as your only alternative is probably a VC (venture capitalist),

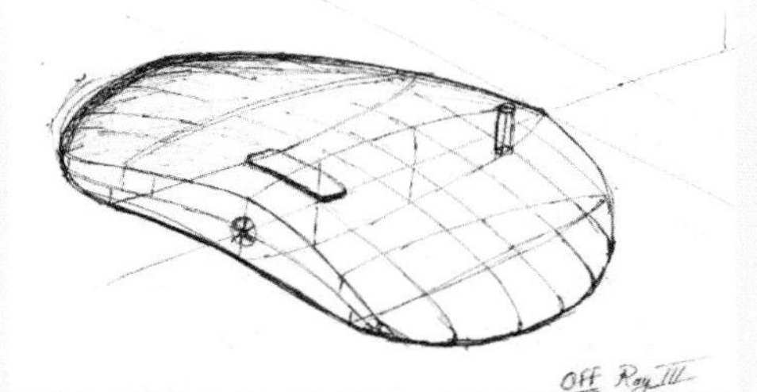

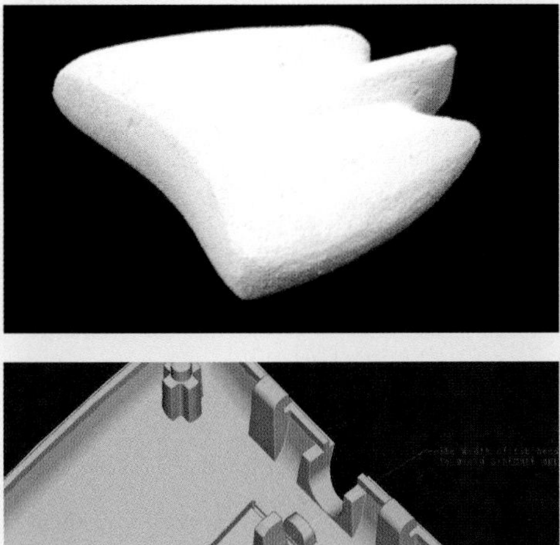

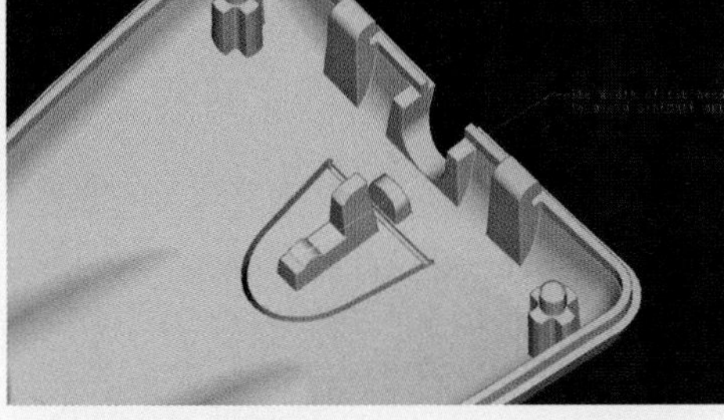

MAKER MADE: From rough concept sketches, to foamcore models, to 3D digital files, the TV-B-Gone Universal Remote moves from an idea to a successful maker-made product.

Photography by Mitch Altman

which often means losing control of your project (and your ability to love it).

3. MANUFACTURE YOUR PRODUCT.
It's much easier to have a life if you hire a contract manufacturer (CM) rather than manufacture your product by yourself. Many CMs can take charge of the process and turn your idea into pallets of packaged products. However, the more aspects of your project you have completed beforehand, the less the CM will cost. Most people go to a CM with a prototype of their project, complete with artwork for packaging, and the CM takes it from there.

It's important for me that TV-B-Gone makes the world a better place, and not at the expense of the people who manufacture it. I used the following criteria in choosing my CM:

» Good quality production
» Treat and pay their employees well
» Have and adhere to safety standards
» Treat the environment well
» Give me a price I can afford

The CM I chose fulfilled these criteria. They had an office in the San Francisco Bay Area and manufacturing facilities in China. Since my criteria are important to me, it was worth a visit to China to interview the people personally and to check out the plant for myself.

Since your project will be manufactured on assembly lines, your CM will create an assembly process that makes sense for your project. Your CM orders the parts and may hire other companies, as needed, to make the electronic boards or make your plastic cases. Once these are ready, the assembly line can begin.

No matter where you manufacture, the cost and time of shipping is a consideration. Shipping companies, called freight forwarders, charge by weight and volume, as well as distance. It turns out that even with import tax and shipping costs, it was much more affordable for me to manufacture outside the United States. Air shipping is the fastest — about three days — but costs about four times as much as surface shipping, which takes about a month.

4. MARKET AND SELL YOUR PRODUCT.
You now have your product manufactured, and shipped to you. So where do you sell it? Stores? Online? What's the retail price? What's the wholesale price?

Pricing — I started out thinking I would only sell TV-B-Gone online. If I sold it for $14.99 (all prices in America end in 99),

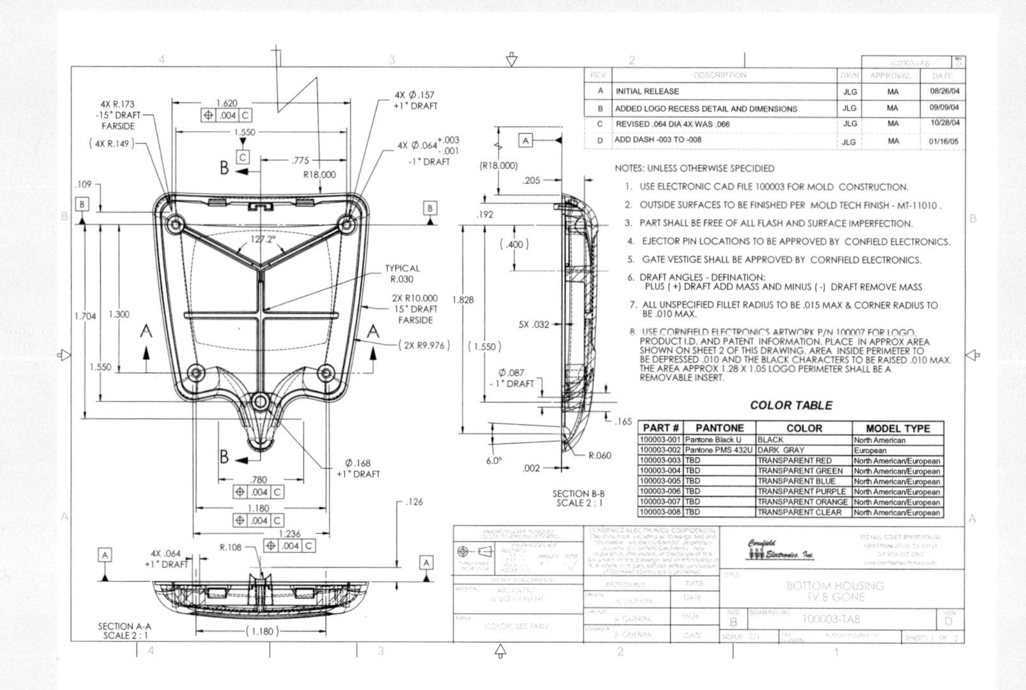

REV	DESCRIPTION	DWN	APPROVAL	DATE
A	INITIAL RELEASE	JLG	MA	08/26/04
B	ADDED LOGO RECESS DETAIL AND DIMENSIONS	JLG	MA	09/09/04
C	REVISED .064 DIA 4X WAS .066	JLG	MA	10/28/04
D	ADD DASH -003 TO -008	JLG	MA	01/16/05

COLOR TABLE

PART #	PANTONE	COLOR	MODEL TYPE
100003-001	Pantone Black U	BLACK	North American
100003-002	Pantone PMS 432U	DARK GRAY	European
100003-003	TBD	TRANSPARENT RED	North American/European
100003-004	TBD	TRANSPARENT GREEN	North American/European
100003-005	TBD	TRANSPARENT BLUE	North American/European
100003-006	TBD	TRANSPARENT PURPLE	North American/European
100003-007	TBD	TRANSPARENT ORANGE	North American/European
100003-008	TBD	TRANSPARENT CLEAR	North American/European

and my cost per unit was a few dollars, then I'd make a bunch of dollars per unit. Seems reasonable. But if you have a popular product, you need lots of help. And you have to pay your help.

And if a store wants to buy your product, then you have to sell to them for about half the retail price. There are many larger stores that prefer to buy from distributors, and the distributors need to make money.

I chose to accept credit cards online, which meant I needed a credit card processing company, who, along with PayPal and Google Checkout, all took a cut.

Oh, and then there's the cost of hiring a company that answers phones for wholesale customers' orders and sends these larger orders to my fulfillment center.

With all my hard work, I was barely breaking even. I had to raise the retail price to $19.99.

Then I learned a lesson in retailing: *the retail price should be set at five to six times your manufacturing cost!*

Website — You need a website, of course. Make it look good! You'll need an online shopping cart to sell your product. These can be purchased, or you can program your own, as I did. Make it easy to navigate to the Buy page, and simple to go through the buying process.

Order fulfillment — When customers place orders, you need to get your product to them. This is not trivial! You could fill a few orders a week yourself. But imagine receiving more than 20,000 orders in the first several weeks, as happened to me. I had a wonderful problem on my hands.

Luckily for me, my friends hired their friends to stuff mailers with TV-B-Gones, label them, print and apply postage, fill out customs forms — it was an international business from the start — and get everything to the post office, working like crazy day and night. This was stressful!

Fortunately, fulfillment houses exist to perform this task. As with your CM, you

CONCEPT TO CREATION: (Opposite) TV-B-Gone design drawing. (Left) Employees at a factory in China assemble the TV-B-Gone. "Assembly lines can be difficult places to work," says the author. "If it's important for you to not add more misery into the world, please choose a CM that treats its people well." (Above) The finished TV-B-Gone, ready for the world.

need to know that your fulfillment house is doing its job. I went through three before I found a match — the right products were getting to the correct customers on time.

Customer support — Be ready to solve problems for people who buy your product. To minimize the amount of support needed, make sure the instructions are plainly visible on the packaging.

You'll probably want to hire someone to answer the flood of emails you'll receive. I have a set of answers to frequently asked questions, all of which are on the FAQ page of the TV-B-Gone website (tvbgone.com).

Publicity — How do you get the word out about your cool product? Advertising is expensive. But there's free advertising that's actually better. It's called "news." Media outlets are always looking for content. Give it to them, and they'll love you. And if one high-profile website or magazine or radio or TV program picks up your story, then others jump on board, and soon the world is talking about your project!

MITCH'S ONE RULE OF DOING BUSINESS

This is the most important paragraph of this entire article. This rule was developed with a considerable amount of pain on my part and I implore you to take it to heart. Here is my one rule: *Only do business with people you like!*

This rule applies to all aspects of doing business. It applies to employees and contractors that you may want to hire. It applies to the people lending — or even giving — you funding. It applies to companies who provide services for you, and to people who want to give you publicity. And it even applies to your customers.

With my TV-B-Gone business I pay all 12 people who help me. I get what remains. It turns out that this is just barely enough for me to live on. And that's great: I make enough money to live a life I love!

1+2+3

Paper Clip Paper Holder

By Gus Dassios

WHAT'S BETTER AT HOLDING PAPER than a paper clip? A bigger paper clip! Here's a very simple project that was completed in about 5 minutes. Finding the parts and tools took longer!

YOU WILL NEED

- **Wire clothes hanger or other stiff wire**
- **Paint and primer (optional)**
- **Wire cutters or hacksaw**
- **Lineman's pliers**
- **File**
- **Electrical tape**
- **Bench vise** or another way of making a clean bend
- **Paper clip (optional)**

1. Cut the wire.

Cut a 15" length of stiff wire. Wire cutters make the cut easy, but a small hacksaw will work, too.

You can use the bottom of a wire clothes hanger, so that you don't have to straighten out any of the bends. Naturally, a longer length will make a larger clip and a shorter length will make a smaller clip.

File the sharp ends until smooth.

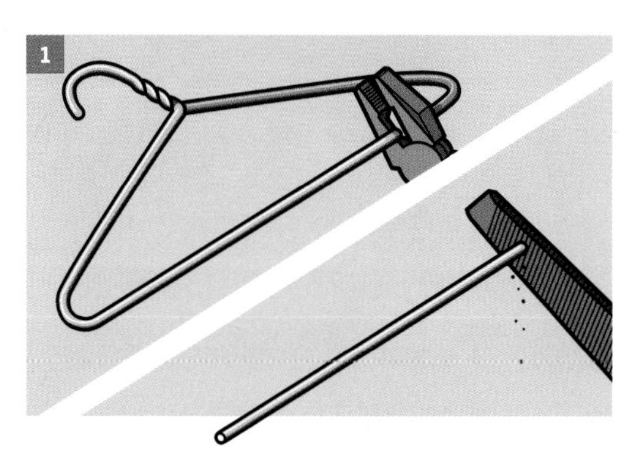

2. Bend it into a paper-clip shape.

Look at a normal paper clip to help you determine where and how to make your bends.

Someone with strong hands should be able to bend the wire without any tools, but lineman's pliers make it easier to twist the wire into a large version of a traditional paper clip.

Wrap some electrical tape around the teeth of the pliers so they don't dig into the wire too much.

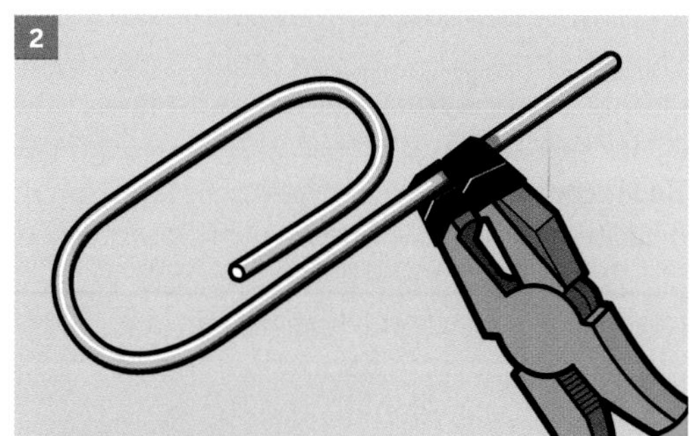

3. Make a 90° bend.

For the final bend — required so that the clip can be freestanding — use a vise. Secure the bottom quarter of the clip in the vise, and then bend the top over into a 90° bend.

After removing the clip from the vise, bend it by hand another 10° farther, so that it won't tip over easily.

If the coating remains intact on the clothes hanger wire, you're finished. If you chipped the coating during the bending process, or if you used bare metal wire, you can prime and paint your paper clip paper holder.

Illustrations by Damien Scogin

Gus Dassios lives, designs, and builds in Toronto, Ontario.

Make:

KARTS and WHEELS

LET'S ROLL!

40 DRILL KART
46 SIMPLE LONGBOARD
48 MOTORIZE YOUR BIKE
50 OUTRUN ARCADE CAR
54 CYCLEKART RACING
56 DOWNWIND CART
60 GRAVITY RACER
65 WILD WHEELED WORLD

Illustration by Juan Leguizamon

Got an itchy trigger finger? How about two? You'll love driving the Drill Kart, an easy-to-build go-kart powered by twin cordless drills that make it steer like a tank.

Educator and MAKE author Gever Tulley helps kids build their own drill-powered go-karts at his Tinkering School in Montara, Calif., using small welded frames, bike parts, and cheap corded drills. We asked him to build a version for this issue, but we wanted it to be large and powerful enough to carry an adult driver, and we wanted to go cordless.

Our kart would need a different geometry from the original kids' karts to accommodate full-sized drivers. This meant changing the drive trains, the wheels, the seat — everything. And when Gever and his assistant Theo Gough arrived at MAKE HQ in January toting buckets of bike parts and steel box beam for welding a frame, we knew it needed another change. There's no welder here at MAKE Labs, so we decided to build our kart out of lumber, not steel.

Together we conceived and built a new Drill Kart. We had a great time working through the design problems while we surmounted each challenge (metal frame to wooden frame, eliminating the front fork, moving the center of gravity), and we think we ended up with a good balance: functional, fun to drive, and easy to build.

We decided on a 3-wheel design with a heavy-duty caster in back and two 20" bike wheels in front, powered via chain drive by two 36V cordless drills. This meant we'd need two more bike hubs chucked into the drills somehow to use as drive gears. Here's how we did it.

MATERIALS

2×4 lumber, about 40' for the frame. Instead, you could use steel box beam and a welder.
Plywood, ½", 10–15 sq. ft. for the deck
Wood screws or deck screws, 2½" (1 box)
Wood screws, 1" (8–12)
16" or 20" bike rear wheels with gear hubs (2) like from kids' bikes or BMX bikes. They're cheaper and give more torque than full-sized bike wheels.
Caster, heavy-duty
Steel brackets, flat, straight, about 1"×5" (4) You can buy mending plates at the hardware store; we made our own by drilling ⅜" holes in flat steel bar.
Bolts, ⅜", with matching nuts and washers: 4½" (4), 3" (4) for the dropout brackets and the caster
Machine screws: 3⁄16"×2½", with nuts and washers (12); 6-32×1½" with nylon insert nuts or pairs of jam nuts (6)
Cordless drills, variable speed (2) preferably 36V. We used 2 Bosch Litheon Brute 36V drills.
Bike rear wheel hubs (2) Freewheel hubs are easily salvaged from damaged wheels (ask a local bike shop). To get a nicer alignment of the drills on the kart, you can substitute one fixed ("fixie") hub, or weld one freewheel to act like a fixie.
Bike drive chains (2)
Seat We got ours from an old office chair.
Steel pipe, 1" diameter, 3'–4'
Optional: Nylon tie-down straps, plexiglass, paint

TOOLS

Saw(s) for crosscutting 2×4s and cutting plywood
Drill and drill bits for wood and metal
Forstner bits, ⅜" and 1"
Driver bits for screws
High-speed rotary tool (Dremel) with cutting wheel
Metal file or grinder
Socket set and ratchet, or various wrenches
Welder (optional)

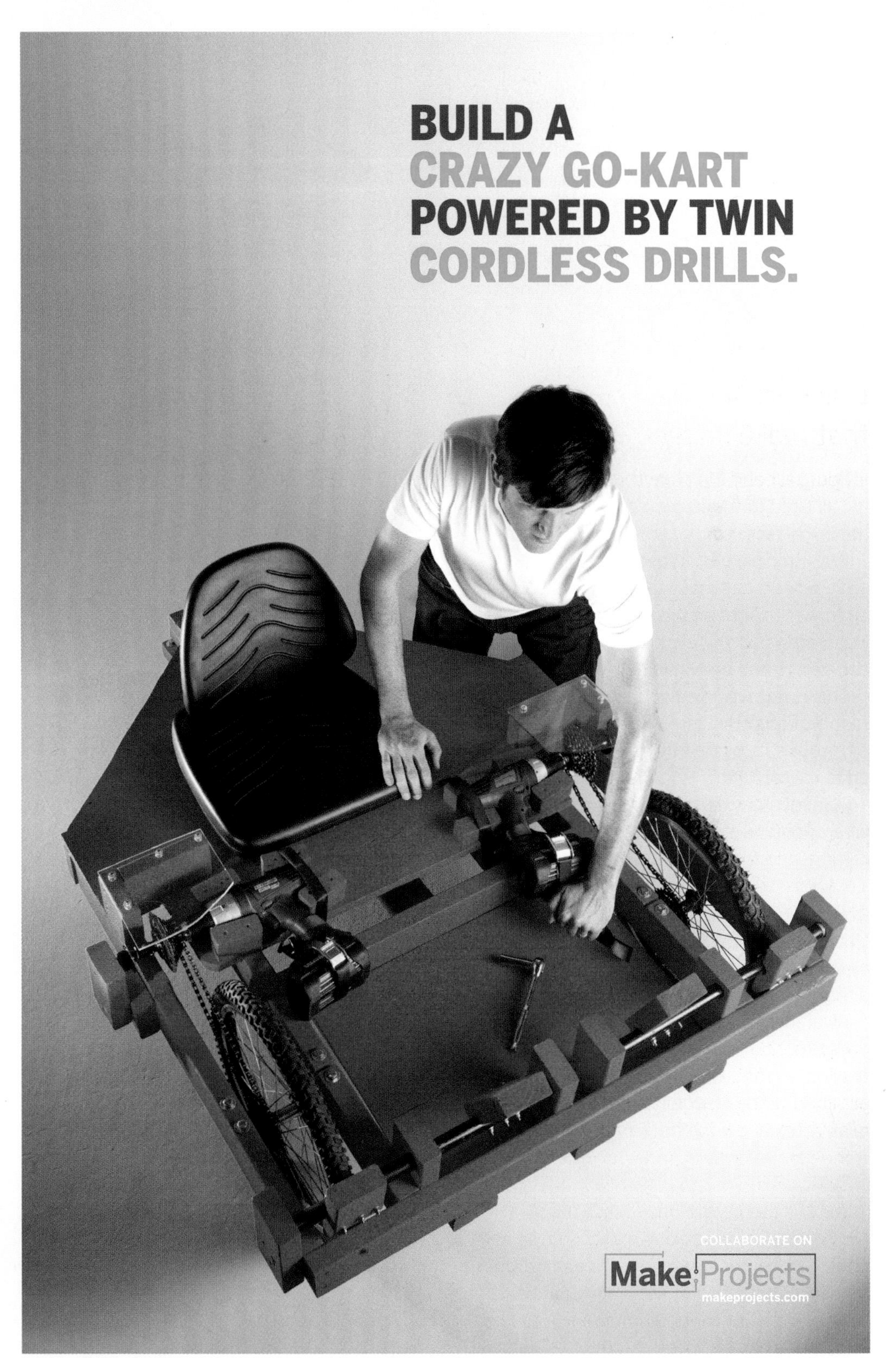

Photograph by Garry McLeod

START

1. DESIGN A FRAME.

Lacking a welder, we began sketching a frame made of 2×4 lumber. For the steering, we originally thought of "Frankensteining" a recovered bicycle front fork onto our wooden frame and putting the 2 drive wheels in back. But the more we looked at it, the more it just seemed heavy and inelegant (Figure A).

So we decided to replace the front fork with a rear caster and bring the 2 paired wheels in front, like on a "tadpole" trike. But they would also be the drive wheels, and you'd steer tank-style by accelerating one motor or the other! We all liked this idea, as it drastically raised the crazy factor of our kart.

We originally envisioned dropping the drive wheels between wooden forks, but the open wood forks on this "box with wings" just weren't strong enough (Figure B). So we decided to box the wheels in (Figure C).

This design also leaves room to install foot brakes at the kart's front edge. You can size your kart by test-fitting the seat and its distance to your imagined foot brakes. Once you've located the seat, you'll know roughly where to mount the drills for comfortable driving, and thus how big to make your frame.

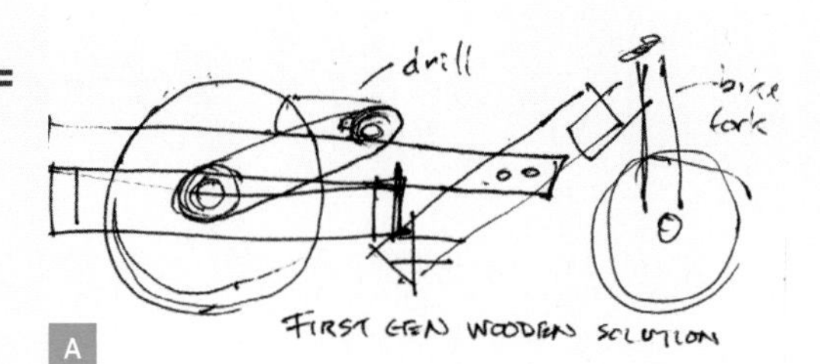

A

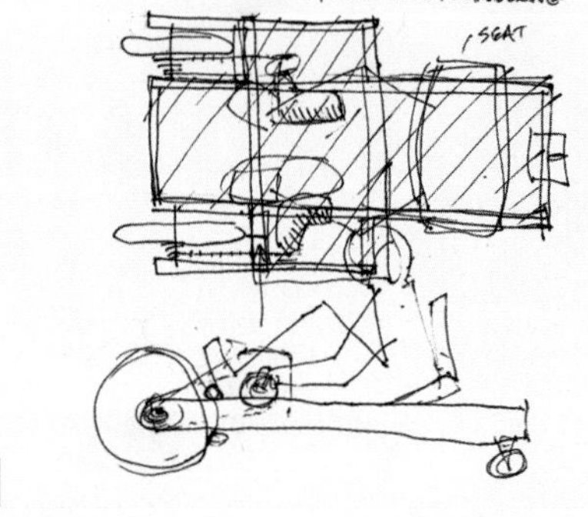

B

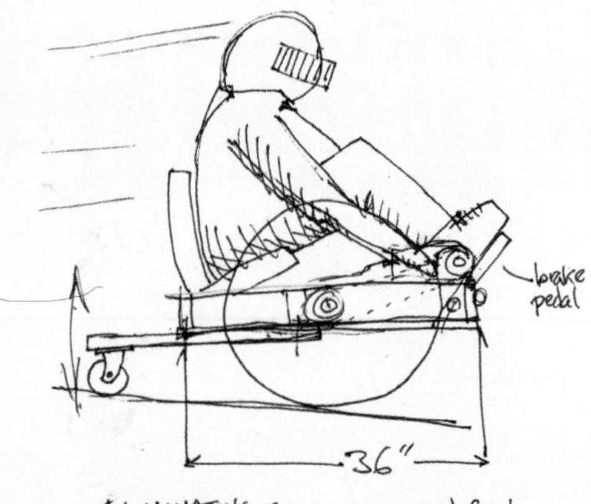

C

2. BUILD THE FRAME.

Our frame is a big rectangle with internal braces to support the wheels and driver, plus a triangular rear extension to stiffen the long central tailpiece that holds the rear caster.

Cut the 2×4s to length, lay them out, and mark where the wheel axles will go. Use the ⅜" Forstner bit to drill 2 dropout grooves for each axle. Screw the frame together with 2½" wood screws or deck screws (Figure D).

3. MOUNT THE WHEELS AND DECK.

Flip the frame upside down. Lay an axle in its dropouts, center a bracket over each dropout, mark and drill the bolt holes through the 2×4, then bolt the brackets on (Figure E). You can unbolt them to remove the wheel.

Bolt the caster to the tail of the frame. You'll probably have to add wood blocks in between to make it level (Figure F).

Finally, cut a ½" plywood deck to cover your frame, and screw it down with 1" wood screws (Figure G).

4. MAKE THE BRAKES.

We settled on simple wooden brakes that rub on the tires, with foot pedals for leverage.

Drawings by Gever Tulley

Photography by Sam Murphy

To make the brake pads and foot pedals, use the 1" Forstner bit to drill round grooves in scraps of 2×4, as shown in Figure H. Use the 3⁄16" machine screws to bolt a brake pad to a length of 1" steel pipe. Bolt a foot pedal at approximately the opposite angle; then mount the pipe in scrap 2×4 supports, so that the brake pad rubs the tire when the driver steps on the pedal (Figure I).

Repeat for the other wheel.

5. ADD A SEAT.
From an old office chair, we got a comfy seat plus a seat back with a post that mounted easily through the frame (Figure J). Your mileage may vary.

6. PREPARE THE DRIVE HUBS.
Liberate the rear hubs and axles from some old bike wheels by cutting the spokes off with a Dremel (Figure K, following page). You'll use

TIP: To prevent the axle from slipping and spinning in the drill chuck, use a file or grinder to create 3 flats on the axle, at equilateral angles (Figure N). Re-chuck the axle, aligning the flats with the faces of the chuck.

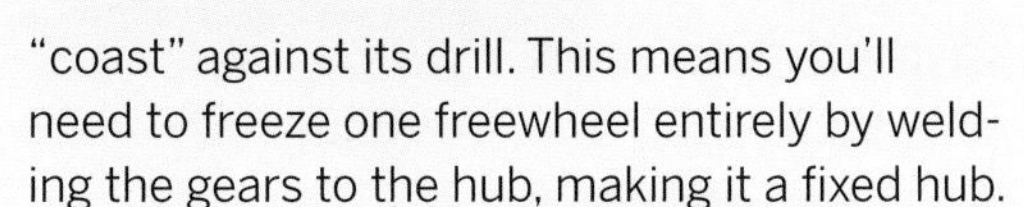

these as the drive hubs, mounted in the drills.

First, you need to lock the axles to the hubs. The easiest way is to jam the axle bearings mechanically. Unscrew the axle nuts and remove the ball bearings (Figure L), then overtighten the axle nuts back down as hard as you can. The axle should now be frozen.

If the axle works loose, fix it permanently by drilling and pinning the hub through the axle (Figure M), using 6-32 machine screws and jam nuts (or nylon insert nuts). We used 3 screws per hub to prevent them shearing off. You could also weld the axle to the hub.

The drills face opposite directions on the kart, but you have your choice of how to align the drive hubs in the drills. If you align both hubs in the same direction, you won't need to do this next modification.

But we aligned our hubs in opposite directions, because that way the drills line up much better for sitting between them. If you do this, then one of your drive hubs must be drivable backward, so that it doesn't just "coast" against its drill. This means you'll need to freeze one freewheel entirely by welding the gears to the hub, making it a fixed hub.

To avoid doing this "fixie" modification, you could also substitute a real fixed rear hub (popular with road racers and bike hipsters), or buy a drive sprocket like we did in the Drill Rod scooter (*MAKE Volume 21, page 108*) and engineer your own axle to mount it in the drill.

Put the axles in the drill chucks and tighten them well.

7. MOUNT THE DRILLS AND CHAINS.

The drills drive 2 bike chains that turn the wheels, and the gear clusters of both wheels face the same side of the cart.

Aligning the drills is tricky — you've got to fix them securely, within reach of the driver's hands, and so that the drive hubs line up with the wheel hubs, without the chain twisting or rubbing on anything. To get the most torque, run the chain from a smaller gear on the drive hub to the largest gear on the wheel hub, like

Photograph by Keith Hammond (Figure K)

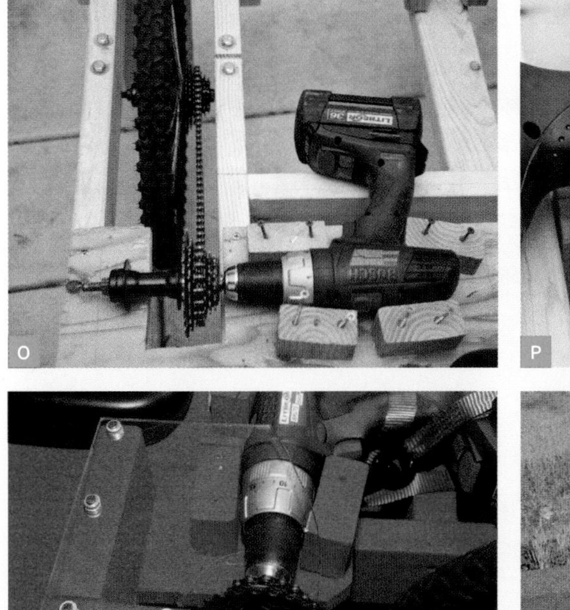

O

P

⚠ WARNINGS:
The drill kart is heavy, so give it a wide berth and take care not to break a leg. (On the upside, all that weight contributes to a nice low center of gravity, which helps prevent tipping.)

Also, take care not to overheat your drills with too much stopping and starting or too steep a hill. Keep an eye (and a nostril) on them; if they're smoking or stinking, give 'em a rest.

Q

R

downshifting to "granny gear" on a bike.

When you've got the alignment and chain tension right, screw down 2×4 blocks to the deck to hold each drill in place (Figure O). We added straps, threaded through the deck, to tighten down the drills but still allow for easy removal (Figure P). You could also box the drills in completely, if you don't want to remove them very often (just remember to leave access to the drill's battery and vent).

To adjust the chain length, you can add or remove links and rejoin the chain with a master link or half-link. To fine-tune the tension, you could even add an idler or tensioner; this could be as simple as a smooth bolt for the chain to run over, or more complex, like the spring arm from a bike's rear derailleur.

8. CUSTOMIZE (OPTIONAL).

We painted our kart a bright, MAKE-logo red, and installed cool see-through guards made from acrylic plexiglass over the drive hubs (Figure Q).

9. MOTOR!

Take it for a spin — and we do mean spin (Figure R). The Drill Kart's crazy 2-trigger tank steering takes some getting used to, and the rear caster loves to spin out. The dual brakes should keep you out of trouble until you get the hang of it, so use them early and often.

Have fun and be careful. You're a motorist!

⊞ Build this kart, share your projects, and see a video of the Drill Kart in action at makezine.com/go/drillkart.

Special thanks to MAKE Labs engineering interns Eric Chu, Brian Melani, Tyler Moskowite, and Nick Raymond for building, refining, and testing the Drill Kart; to Robert Bosch Tool Corp., and to West Valley Welding in Sebastopol, Calif.

Gever Tulley is co-founder of Brightworks, a new K–12 school in San Francisco (sfbrightworks.org).

Last year my 13-year-old daughter asked for a skateboard, so I gave her my 20-year-old board, which wasn't seeing much use. She liked to ride it along the bank of the Los Angeles River, and I would ride with her on a Razor scooter. But after a while, I started to miss having a skateboard, and I thought it would be fun to make one.

I put it off for a few months, until I saw a video of Lloyd Kahn, the well-known maker and former geodesic dome guru, cruising down a gently sloped street (makezine.com/go/kahn). He's 75 years old, and his skating skills are top-notch. This was the inspiration I needed to get off my butt and make my own longboard.

I went online to look for plans. There are many plans and kits available, but they seemed overly complex for what I was setting out to do. I simply wanted a 4-foot-long board that wouldn't sag too much when I stood in the middle of it. My solution was to make a board with a hump in it.

Here's how I did it.

1. Glue. I squeezed a liberal amount of Gorilla Wood Glue on one side of two 48"×8" lengths of ¼" Baltic birch plywood. Then I used a paintbrush to spread the glue in an even layer. I stuck the glued sides of the 2 boards together.

2. Bend. I suspended the board between 2 bricks placed at its far ends. I set a couple of heavy boxes of books in the middle of the board so that it sagged in a U shape (Figure B).

3. Clamp. Using every clamp I could find around the house, I pinched the boards together. I did not disturb the setup for 24 hours.

The next day when I removed the boxes of books, the 2 pieces of wood stayed bent. I flipped the boards over, stood on the hump and bounced up and down a bunch of times to make sure it could support my weight without breaking. It passed the test.

4. Mark. I drew a template of the end of the skateboard using Adobe Illustrator, cut it out, and taped it to one end of my board. Then I used a pencil to transfer the shape to the board. I used the same template on both ends.

5. Cut. I cut out the rough shape of the skateboard using a band saw (Figure C). Then I sanded the edges.

6. Add wheels. I bought a set of trucks and wheels from Amazon for about $35. I installed them on the board and tested it out in my driveway. I discovered that when I leaned into the skateboard, the wheels came in contact with the wood. The skateboard would come to a screeching halt, sending me sprawling.

Photography by Mark Frauenfelder and Linda Nguyen (Figures A and E)

A

MAKE A TERRIFIC SKATEBOARD WITH A MINIMUM OF FUSS OR TOOLS.

You can make it!

B

C

D

E

So I used the band saw to cut clearance arcs for the wheels (Figure D). This did the trick. Now I could make turns without having to worry about the wheels jamming into the wood.

7. Finish. I painted a design on the bottom of the skateboard and sprayed on several coats of polyurethane.

I added clear grip tape to the entire top surface of the skateboard. (I found out that I don't like clear grip tape, because it gets dirty really fast. Next time I'll use black grip tape.)

After installing the trucks and wheels, I invited my daughter to go skateboarding with me along the bank of the L.A. River again. She took her skateboard and I took my new one. My board worked beautifully. I couldn't have been more pleased with the way it handled. My daughter asked if she could try it out.

"I love this!" she said, as she glided smoothly down the paved embankment. "Can we trade?"

It looks like I'll be making another long-board soon.

Mark Frauenfelder is editor-in-chief of MAKE.

I was reading Boing Boing one day and saw a link to bikemotorkit.com. At that site, Gas Imports of Kalamazoo, Mich., was offering a bicycle motor kit consisting of a 66 cubic-centimeter two-cycle engine, drive mechanism, gas tank, muffler, and controls for $120.

The ad brought back kid-time memories for me, as I had dreamed of one day owning a Whizzer — now I could make one!

THE WHIZZER STORY

Whizzer motorbike kits were introduced in 1939 by Breene-Taylor Engineering, a Los Angeles producer of aircraft parts and carburetors. The kit cost $55 (something like $800 today). People would buy these kits, which consisted of an engine, roller drive, control levers, etc., and adapt them to regular bikes. (Interesting trivia: a Whizzer was the only motorized vehicle you could purchase new during World War II!)

MAKING A WHIZZER KNOCKOFF

The Whizzer brand was revived in 1997, and today a new Whizzer sells for $1,400; the engine kit alone is $500. I decided to pay $120 for the Gas Imports kit and maybe $50 for a used mountain bike, and make my knockoff motorbike for a lot less money.

The kit arrived from China about three weeks after I ordered it. I opened the box to find that every nut, bolt, washer, and other sundry parts were not packed in tidy little plastic bags but were loose and had obviously rolled around in the box during its journey to me. I had no idea if parts were missing, as there was no assembly manual either.

TWO YEARS IN THE MAKING!

I carefully laid out the parts on our laundry room counter and forgot about the mess — for two years. After this lapse, I figured I'd better get serious about building my motorbike, especially since my wife wasn't going to put up any longer with lost counter space.

I went online and found an assembly manual PDF and an assembly video at Gas Bike (gasbike.net). They also sell a variety of bike motor kits. From there, I went to Craigslist and spotted a mountain bike for $35. The bike was in pretty rough shape, so I overhauled it first before motorizing it.

Now for the coup de grâce — the engine and control mounting! Actually, the process went quite well. The kit uses a chain-driven sprocket that clamps to the rear wheel's spokes. I found I was missing a part and readily located a replacement at a scooter shop.

INAUGURAL RIDE

Since the now-motorized bike has no starter, I had to get on, aim it downhill, let out the clutch, and hope it started. It did! A nice pop-pop-pop noise like a baby Harley came from the muffler. I turned it uphill and was amazed at the torque that came from that tiny engine.

I'm looking forward to more rides. My only regret is missing two years of fun motorbike riding by letting the parts lie fallow instead of getting with it the day the kit arrived!

+ Bike engine kit forum: motorbicycling.com

Lew Frauenfelder lives in Boulder, Colo. When he's not building bikes, he's making stained glass windows, fixing stuff around the house, and trying to stay retired.

Photograph by Garry McLeod

ADD A RETRO-STYLE, ONE-CYLINDER GASOLINE ENGINE TO YOUR BIKE.
SD Stinger
1-888-KIT-BIKE

Photography by Four Eyes Photography, Art Center College of Design

I consumed many hours as a teenager in arcades playing classic video games, and spent a small fortune in quarters mastering a few of them. One of those games was *OutRun*, a driving game released by Sega in 1986, which featured a red Ferrari Testarossa racing down a freeway that snaked through a variety of landscapes.

At the time of its release, this game had a number of interesting features: a steering wheel controller that actually shook when you hit the ditch, user-selectable "chiptune"-ish soundtracks, multiple different endings, and a graphics processor that created an immersive sense of speed — or at least immersive enough to convince a 14-year-old to spend his paper-route money.

Twenty years later, I came across this game at the Beach Boardwalk arcade in Santa Cruz, Calif. It wasn't in the standard upright "fridge format"; this was an 800-pound, car-shaped, sit-down cabinet with fiberglass wheels, working taillights, dashboard, and a powerful hydraulic system that shook the entire cabinet from side to side when you veered into the ditch.

The cabinet was as big as a golf cart, and it got me thinking: "What would it be like if this

WORLDS COLLIDE: (Opposite) Brody Condon drives the *OutRun* vehicle at the Zer01 Biennial in San Jose, Calif., in September 2010. The *OutRun* project explores the overlap between the physical world and game environments by combining a real-world vehicle and *OutRun*, an 8-bit arcade driving game released by Sega in 1986.

Fig. A: Disassembling the original OutRun arcade game cabinet (background) and three-wheeled 1959 Turf Rider Mark IV golf cart (foreground).

could actually roll down the street?"

In my mind, this would be something like people driving while blindly following their GPS vehicle navigation systems and getting into accidents — or some form of an augmented reality video game concept car. Either way, I liked the idea. It reminded me of the type of dream cars that Ed "Big Daddy" Roth had built: souped-up hearses, hot-rodded kids' Radio Flyer wagons, and flying-saucer cars.

SKETCHING IT OUT

I thought the idea over and decided to go ahead with it in fall 2008. The design process began with mocking up a general sketch in Photoshop. Using some photos I'd taken of the game in Santa Cruz, I put together an orthographic view of a three-wheeled cabinet and placed it on a road scene.

In order for this project to move beyond an extreme case mod, I'd need to display something more than the original arcade game on the screen in front of the driver. Ideally, I envisioned that the screen would transform the real world into an 8-bit video game: in other words, I thought the system should make the entire world as its playing field. I sketched out the idea of making an 8-bit-looking "skin" for a GPS navigation system.

I wasn't sure how the system would work, but I located a nonfunctioning *OutRun* cabinet in Maine through an online discussion forum and had it shipped to me in California.

Armed with my Photoshopped images and a PayPal receipt from a stranger, I pitched the idea on the last day of class before summer break to my first-year undergrad video game development class at UC Irvine. I said that if anybody wanted to help me out with the project over the summer for free, I'd be happy to exploit their energy and talent. To my surprise, five students volunteered: Chris Guevara, David Dinh, Matt Wong, Erik Olson, and Richard Vu. We met and figured out tasks: studying the original game, looking into GPS software, locating a suitable drivetrain.

PUTTING MY CART BEFORE MY CART

The arcade cabinet arrived and only had minor damage. I found and bought an inexpensive three-wheeled Turf Rider golf cart from 1959 on Craigslist that looked like it came from *The Jetsons* and had similar measurements to the cabinet (Figure A).

After disassembling the cabinet and cart, we came to the conclusion that our golf cart wouldn't work for a number of reasons: its single front wheel wouldn't provide enough stability to prevent the car from tipping over, and its treadle steering would be difficult to mechanically couple to the faux-Ferrari arcade steering wheel. In order to maintain a low seat height and the original look of the arcade game, we'd have to reposition all six batteries from their original location under the golf cart seat: three would fit behind the rear axle, while three would need to be put in a top-heavy

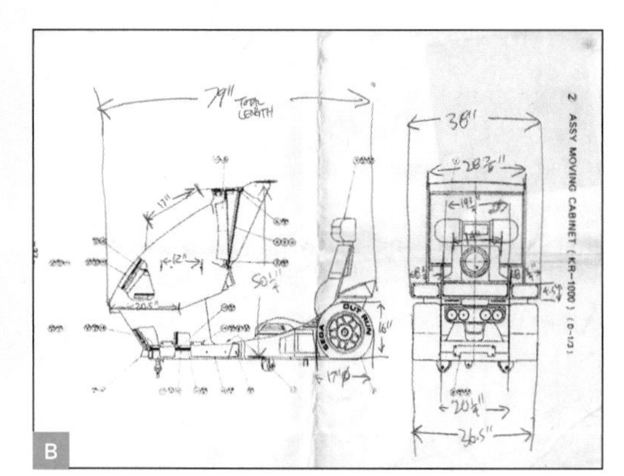

Fig. B: Proposed component layout sketch for the system. We later figured out that half of the golf cart batteries would need to be positioned behind the rear axle.

Fig. C: Welding the custom frame, which combined components of the golf cart and arcade frames.

Fig. D: Two-dimensional, 8-bit-style media assets assembled using Google SketchUp to re-create the start sequence of the original *OutRun* game on Park Avenue, Balboa Island, Calif. This scene was tested in real time with a GPS receiver and Google Earth.

Fig. E: (Opposite) Alex Szeto works on the integration of hardware, software, sensors, and physical components.

position above the steering column (Figure B).

The project caught the interest of the Center for Computer Games and Virtual Worlds at UC Irvine. They asked me to work full time on it. They also provided staff support and a modern golf cart drivetrain with rims and tires.

In a grueling push over the summer of 2010, the physical drivetrain was pieced together. Although it ended up looking similar to how it started, it was custom-built from the inside out: frame, steering column, wiring, power management, sound system, sensors, brake lines, and throttle pedal. The original mahogany cabinet and the golf cart were Sawzalled, welded, and drilled into submission in more than 1,000 hours of sweat and bloody knuckles (Figure C).

RETHINKING THE DRIVER INTERFACE

The software development of the system was not straightforward. We thought that a GPS system would be the best solution, but we soon learned that many locations where we wanted to drive didn't have street data, the cart's driving speed was painfully slow when rendered in the style of a video game, and the resolution of most consumer-grade GPS systems was too low for what we needed. We came to these conclusions after building and testing a prototype system that featured 8-bit game sprites and roads positioned into Google Earth (Figure D).

Using a camera-based computer vision system seemed like a better approach. I contacted my friend Jeremy Bailey, who provided a quick mockup in Max/MSP/Jitter to give me an idea of how this could be accomplished.

My student Chris Guevara figured out how to visually detect real-world features through a single camera system. Our first main task was to find the vanishing point of real-world roads, since they were the most important feature from the original video game.

We used real-time edge detection and custom filters to identify what we thought were

road-like curb lines. We then calculated the average vanishing point of these lines, and as long as we had a clear and open road with a painted curb and few shadows, the road detection system worked well. The system also worked while looking down hallways, and in unexpected situations, like making a pyramid shape with your arms above your head.

Although the vision system had several inaccuracies, it was considerably more flexible and playful than the location-based GPS system. We built our own version of the original arcade driving game in Flash, and used our calculated computer-vision road endpoint to change the shape of our Flash-based game road (Figure E).

VIRTUAL REALITY ON WHEELS

Making physical things is often a battle with tools and materials, and this project was complicated by trying to integrate computer vision, physical computing, custom software, and other systems — like a golf cart.

However, actually making a working real-world system has a significant power of legibility and reality over just mocking up something in Photoshop. Rolling down the street, the arcade cabinet car is understandable and real to a diverse audience — grandmothers and infants, for example — in a way that my Photoshopped image wasn't. And the process of wrestling between the materials and the concept had its own imperfect, *wabi-sabi* beauty to it.

The *OutRun* project was never intended to seamlessly turn a video game into real life and real life into a video game. Instead, it was intentionally built as a type of *chindogu* or critical design — a system that uses paradox, irony, and physical prototyping to raise a series of questions in a provocative way.

In the case of *OutRun*, when the game is extended beyond its normal constraints as a video game, its playfulness malfunctions: it's like a fantasy taken too far, and it results in confusion, nervous humor, and a questionable sense of over-reliance on technology.

To build in this way is to slip into the role of a trickster: using humor and paradox to challenge, bend, and break common assumptions about our everyday lives with technology.

Videos and build notes: conceptlab.com/outrun

Thanks to Walt Scacchi, Chris Guevara, Alex Szeto, Paul Dourish, Gillian Hayes, Jong Weon Lee, Eric Mesple, Yuzo Kanomata, Jeremy Bailey, David Dinh, Matt Shigekawa, Jesse Joseph, Mike Tang, Matt Wong, Erik Olson, Richard Vu, Kari Nies, and Craig Brown for their development efforts and assistance.

Support for this project was provided by the following organizations at the University of California Irvine: the Laboratory for Ubiquitous Computing and Interaction, the Arts Computation Engineering Program, the Center for Computer Games and Virtual Worlds, the Institute for Software Research, and the California Institute for Telecommunications and Information Technology. Support for this research is also provided through grants from the National Science Foundation (#0808783) and the Digital Industry Promotion Agency, Global Research and Development Collaboration Center, Daegu, South Korea. No review, approval, or endorsement implied.

Garnet Hertz is a research scientist and artist-in-residence in the Department of Informatics at the University of California Irvine and is adjunct faculty in the Media Design Program at the Art Center College of Design in Pasadena, Calif. conceptlab.com

Cyclekarts are pint-sized (yet adult-piloted), prewar-style racers that are equal parts art object and four-wheeled mayhem generator. They were developed about 15 years ago by a few enthusiasts in Central California as a way of injecting a little levity into the status- and competition-hungry automobile world.

With 17" Honda motorcycle wheels, a 200cc engine, a lightweight plywood body, and a maximum length of 98", cyclekarts are optimized for fun. They can't be bought — they are built by their owners. The ride is hair-raising, they go anywhere but in a straight line, and they leave their drivers caked in dust and grinning ear to ear. Welcome to the dangerous yet charming world of cyclekarts.

I tracked down Ryan Wasson of Anvil Customs in Springfield, Mo., who has built a few of these machines.

Nik Schulz: When did you get started building cyclekarts?

Ryan Wasson: Just a couple of years ago in 2007. I'm kind of a gearhead and I wanted to build a go-kart for adults. I searched on the internet and found cyclekarts. These guys were holding races in these little four-wheeled wooden karts powered by 6½-horsepower motors.

NS: What are they like to drive?

RW: Ha, they're not very good to drive! You're on a rigid 3"×1" square tubing frame that's basically a box, with a rear live axle that's hard-mounted into the frame. There's no rear suspension, and there's not enough weight to the vehicle to make the front suspension function, although it looks killer.

NS: How long does it take to make one?

RW: If you have all of the parts, it takes about a week. The problem is obtaining the correct wheels. The 17" wheel comes from late-60s, early-70s motorcycles. The size of the wheel is incredibly critical to how the car's going to function. If you get wheels any larger than 17", you won't be able to control it.

The rest of the parts you get from go-kart supply houses. Some of the parts you have to manipulate to make them work, like the rear hubs, so that you can adapt the early model Honda wheel to the go-kart axle.

NS: Where do you find the wheels?

RW: That's a good question. There's no company that manufactures them. You can find them on eBay for $100 to $150 apiece.

NS: What's the biggest challenge to making a cyclekart?

RW: Finding the wheels, but a close second is building it to be substantial enough that it doesn't rattle apart underneath you when you take it down your driveway. There are specs that you need to go by if you're going to build a true cyclekart. The weight of the finished product can't be over 250 pounds.

The front end is probably the most difficult thing to hammer down. You're dealing with two 17" wheels on go-kart hardware that's built to control 4"–6" go-kart wheels. That's

RETRO ROCKETS:

The almost-finished Brannan Special cyclekart, inspired by the Schuco Mercedes tin wind-up toy, and named after a fictional racing hero. It was built in 2006 by Peter Stevenson, originator of the cyclekart concept.

The Type 59 Bugatti cyclekart at speed. Built by Peter Stevenson in 2001, it's got an aluminum body with functional friction shock absorbers. As with all cyclekarts, top speed is 35–40mph, but you'd be very foolhardy to go that fast!

why they're just for fun.

NS: How did you overcome that challenge?

RW: Just take your time, get your welds right, use good quality hardware.

NS: Was it also trial and error?

RW: Oh my god, yeah! Every time I build one it's trial and error! Absolutely.

NS: Do you know if there are pockets of people around the country that are getting together and having cyclekart events?

RW: There's a group down in Texas. Other than that I don't know anyone that knows anything about these, to be honest. It's a real small sect.

NS: Are you still selling frames?

RW: Yes, I sell frames. I also build and sell the whole front end with the drop axle with the welding brackets, spindles, and suspension. So I can sell the frame with the front end and the rear live axle installed, plus the bearings and bearing hangers.

» Introduction to cyclekarting: cyclekarts.com
» Contact Ryan Wasson: anvilcustoms.com

Nik Schulz is a Sonoma County-based illustrator (l-dopa.com), writer, and co-founder of the travel-adventure blog West County Explorer's Club (wcxc.wordpress.com). His work can also be seen in the book *Handy Dad: 25 Awesome Projects for Dads and Kids* (Chronicle Books). He is enamored of vehicles of all kinds and has a penchant for DIY.

Photography by Michael Stevenson

FASTER THAN THE WIND

BY ERIC CHU

MIND-BENDING PROPELLER CART OUTRUNS THE WIND THAT POWERS IT.

Is it possible to build a wind-powered vehicle that travels directly downwind, faster than the wind, continuously? Rick Cavallaro posed this question as a brainteaser on an internet forum in 2006. His own vector analysis showed that it was possible, but the question went viral, with naysayers expressing doubt on numerous online forums and even in the pages of this magazine.

After years of controversy, Cavallaro and his friend John Borton decided to build a cart that would put all doubts to rest. Drawing design ideas from R/C helicopter enthusiast Mark Conroy, they built a downwind propeller cart and documented it in YouTube videos.

I followed their instructions to build my own version, making some changes to the front axle, and it runs like the wind. (Faster, actually.) Here's how I did it.

Illustration by James Provost

MATERIALS

Bearings: 5×13×4 Revolution (2) and 5×11×4 Revolution (3) #695-RSZ and #MR115-RSZ from Avid RC (avidrc.com)
Tail gear set for Century Hawk R/C helicopter #HI3075 from Century Helicopter Products (centuryheli.com) or #LXMJX7 from Tower Hobbies (towerhobbies.com)
Tail gearbox for Century Hawk or Heli-Max Kinetic 50 R/C helicopter Century #HI3078 or Tower #HMXE9914
GWS 3.35" Shock Absorbing Wheels for R/C aircraft (2) Tower #WH01/85
Dubro Micro Lite Wheels, 2" Tower #200ML. They're sold in pairs, but you only need one.
Carbon fiber tubes, 5mm OD: 4mm ID × at least 25" long (1) and 3mm ID × at least 7½" long (1) #CF4 and #CF3 from HobbyKing (hobbyking.com).
Aluminum tube, ⅛" OD × 0.014" wall thickness, at least 1½" long
Music wire (aka piano wire), 0.062" diameter, 1⅛" length Tower #LXWV00
Propeller, GWS EP 381mm×191mm HobbyKing #RD-1575
Cyanoacrylate glue (super glue), foam safe such as HobbyKing #CA460
HDPE bar, ½" thick, at least 1"×1½" order as Plastic Cutting Board from Tap Plastics (tapplastics.com). 1"×1½" is the final size, but ordering 3"×1½" will make it easier to clamp in a drill press.
Aluminum tube, ⅜"×25" from a hardware store
Machine bolts 6-32, 1¼" (2)
Nuts, 6-32 (3)

TOOLS

Drill press with clamps
Dremel (or other rotary tool) with cutting disc
Drill bits: ⅛", ⅜", 1/16", 5/64", #9 or 5mm, 7/16" or 11mm, 5/32"
Hacksaw for HDPE and aluminum tube
Wire cutters and long-nose pliers
Precision ruler (standard) or calipers
Bench vise
X-Acto knife
2.5mm hex key (aka Allen wrench)
Needle or pin the thinner the better
Masking tape
Sandpaper: 100, 150, and 600 grit
Acetone, in acetone-proof dish for soaking
Paper towel
Powered treadmill (optional)

+ Visit makezine.com/26/downwind **to read Rick Cavallaro's article "The Little Cart That Did," which details the development of this small cart and the 23-foot-tall Blackbird, which was officially clocked at 2.8 times wind speed, directly downwind.**

Photography by Sam Murphy

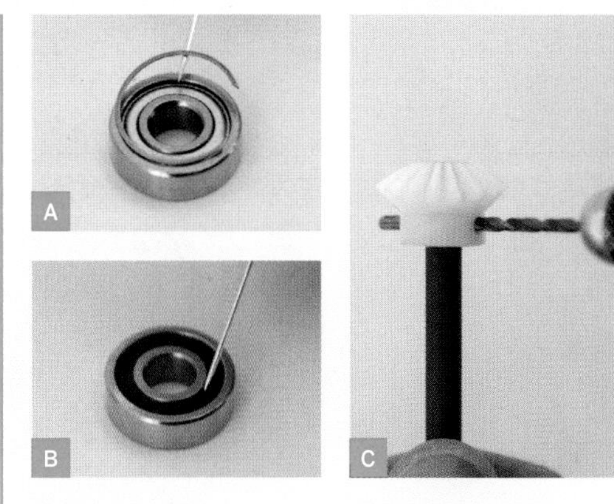

TIPS: To mark locations for cutting and drilling on smooth materials, such as carbon fiber, HDPE, and aluminum, stick on some masking tape and mark on the tape.

Don't sand or drill near the bearings, gearbox, or gears — the resulting dust will hinder the cart's performance.

START

1. CLEAN THE BEARINGS.

Use a pin to remove the C-clip from the side with the metal shield (Figure A). The rubber seal can be removed by prying its inner diameter (Figure B). Soak the pieces in acetone, swirling occasionally. Dry them on a paper towel until they don't smell of acetone anymore.

2. MAKE THE PROP SHAFT.

Use a Dremel to cut a 25" length of 5mm OD × 4mm ID carbon fiber (CF) tube. Use 150-grit sandpaper to sand 5" down one end, and wipe away any dust with a wet towel.

Try to slide the two 5×13×4 bearings down the sanded end of the shaft. If they don't fit, sand and wipe clean until they do.

Remove the bearings to someplace dust-free. Use 100-grit sandpaper to sand the shoulder off the backs of the 2 tail gears.

Fit the larger gear flush onto the prop shaft's end, teeth facing outward. Use a 5/64" bit to drill into one of the gear's mounting holes and through one side of the tube. Align the gear with the drilled hole, and use pliers to push one of the included metal pins through the hole and into the shaft. Drill through the hole and shaft on the other side (Figure C).

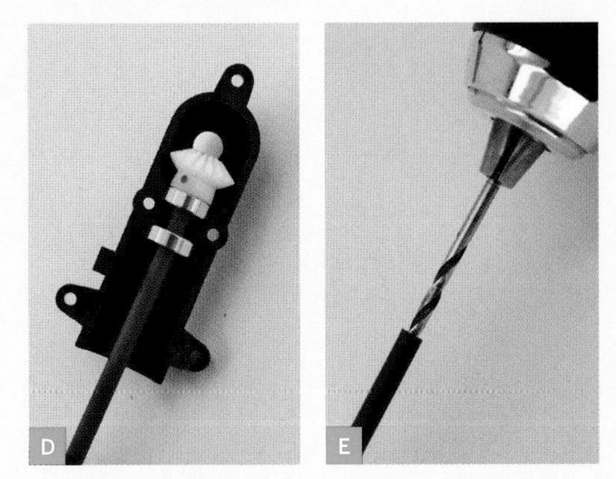

Remove the gear, clean the shaft, and thread the bearings back onto the prop shaft. Replace the gear, fitting the pin through both holes.

Test-fit the prop shaft, gear, and bearings, lengthwise inside one half of the gearbox case (Figure D). To glue the rear bearing to the shaft, slide it down the shaft, apply a thin layer of super glue, slide it back, and replace the shaft assembly in the case before the glue dries. Do not get glue into the bearings.

Fit the other half of the case over the shaft assembly and make sure the gear spins freely.

3. MAKE THE AXLE.

Cut a 7½" length of the 3mm ID CF tube and drill it out with a ⅛" bit (Figure E). Cut two ¾" lengths of ⅛" aluminum tubing and glue them into the CF tube, leaving ¼" exposed on each side. Use the ⅛" bit to drill out the hub holes of the 3.35" wheels so they fit onto the axle.

Mark the center of the 7½" axle with the edge of a piece of tape, and slide the smaller gear over the axle until its teeth lie against the tape at the center. Use the same procedure as in Step 2 to attach the smaller gear to the axle in this location.

Slide one 5×11×4 bearing onto the axle (Figure F). It should fit snugly. If it's too loose, thicken the axle with a layer of super glue; if it's too tight, sand the axle down with 600-grit sandpaper. Remove tape.

4. ASSEMBLE THE GEARBOX.

Insert the axle bearing into the deeper half of the gearbox case, axle running through. Fit in the prop shaft so the gears mesh (Figure G). Fit the second 5×11×4 bearing into the other case half, and close the case. The prop shaft should spin the axle freely. Take the small bolt and nut that come with the gearbox and screw the halves together at the hole near the axle using a 2.5mm hex key.

Identify the gearbox's top by turning the prop shaft clockwise, viewing from the back. The axle should run the wheels (when they're on) away from the propeller, not toward it. Tighten one of the gearbox's longer included bolts through its middle top hole (Figure H).

5. MAKE THE BEARING BLOCK.

The prop shaft is steadied by a block of HDPE that caps the back end of the frame tube. The shaft runs through a 7/16" (or 11mm) hole centered ½" below the block's top 1"×½" face, and the frame tube fits into a ⅜" hole drilled ¾" deep in the center of the bottom face.

If you start with a larger piece of HDPE, you can cut the block to size before drilling, or leave some extra length to clamp in a drill press for drilling the 7/16" (or 11mm) hole. After cutting and drilling, round the block's corners.

6. MAKE THE FRAME.

Cut 25" of ⅜" OD aluminum tube and drill two 5/32" holes through, ¼" and 1.45" (or 1 29/64") from one end. This is the frame tube.

Use a 5/32" bit to enlarge the gearbox's two bottom mounting holes. Slide a 6-32×1¼" bolt through each, and screw a nut onto the front bolt as a spacer. Secure the tube onto

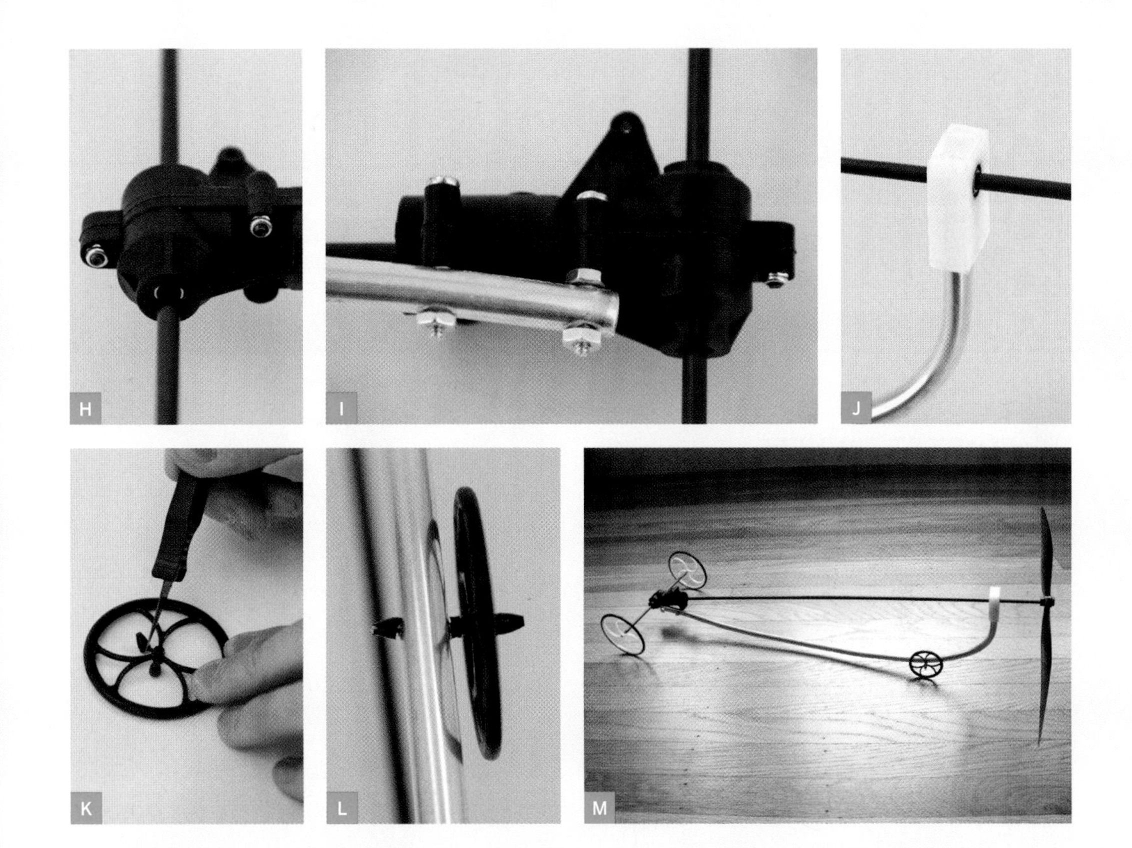

the bolts using 2 more nuts (Figure I).

Bend the frame tube so that its back end T's into the prop shaft about 2" from the back. Slide the bearing block over the prop shaft and fit it onto the frame. Adjust the frame until the shaft does not touch the bearing block. Slide the last 11mm bearing over the shaft and into the block (Figure J). If the bearing doesn't fit onto the shaft easily, sand with 600-grit sandpaper down to where the bearing will rest in the block. Adjust the tube's bend so that the block applies no force to the prop shaft.

7. ATTACH THE DRIVE WHEELS AND PROPELLER.

Fit the 3.35" wheels over the axle; they may need some super glue. Use a #9 or 5mm bit to drill a hole through the center of the propeller, and press-fit the prop onto its shaft.

8. ATTACH THE REAR WHEEL.

Drill a 1⁄16" hole horizontally through the lowest point of the frame tube. Cut 1" of .062" music wire, which you'll use to make the rear axle.

Use a knife to cut the molded hubs off a 2" wheel (Figure K). Enlarge the wheel's hole with a 5⁄64" bit, and use a 1⁄16" bit to enlarge the holes in the hub pieces. Slice one hub piece to turn it into a shorter hub and a spacer.

Glue the long hub onto one end of the wire, then slide the other end through the wheel, spacer, and frame, in that order. Finish by gluing the short hub onto the other end (Figure L). You've made your cart (Figure M)!

You can use a treadmill to adjust the cart's alignment. Hold the cart by the gearbox, lower it onto the belt, and release it after the prop comes up to speed. If it veers one way, bend the frame to compensate. If it drifts backward, flip the prop around (it's unidirectional).

+ See makezine.com/26/downwind for background and sources.

Yo-yo enthusiast and MAKE engineering intern Eric Chu likes to play with robots, come up with crazy ideas, and eat sushi.

WEEKEND WARRIOR

BY JEREMY ASHINGHURST

CONTEST WINNER!
Jeremy's Weekend Warrior gravity racer was the winning entry in makeprojects.com's Karts & Wheels Contest. Complete instructions for building this rad racer can be found at makezine.com/go/warrior.

Photography by Jeremy Ashinghurst

A few months after graduating from college and starting work, I decided I needed a new hobby. Soapbox racing sounded interesting — very fun, not too expensive, and it would get me into metalworking, which I had no experience with as an electrical engineer. I soon found the local soapbox drivers club, the Maryland MISFITS (downhillmisfits.com), and it was, as they say, all downhill from there.

I joined the MISFITS in February 2009 and had big plans for building a long, fast, full-suspension kart, despite having never worked with metal before. The group's organizer, Fran Honeywell, wisely told me to keep those plans in my head, and meanwhile I could start helping him measure, cut, drill, and sand some of our loaner karts. He taught me how to weld, and as a first project, he helped me design and build the Punishing One, a cart I rode to victory in the 2009 East Coast Challenge.

I'm not as competitive as some, however, and I wanted a smaller, lighter, better-handling, and more elegant cart just to enjoy driving

A

on the hills rather than racing. With some more help from Fran, I started designing the Weekend Warrior, the kart described here.

The Weekend Warrior is a gravity-powered vehicle that weighs less than 100lbs and is capable of highway speeds. I chose to give it a full suspension, which is more comfortable and cooler than a lighter, simpler, and cheaper rigid frame. In my experience, it's a wash between the options of handling and speed as long as the suspension is precise and well-made.

During the approximately 300 hours (over the course of one year) it took me to build this kart, I learned the MIG welder, hydraulic tube bender, metal lathe, and tap and die set. And I succeeded in my design objectives. I love my Weekend Warrior; it is certainly my favorite kart, and it made the summer of 2010 my "Summer of Soapbox."

I would drop the kart off at the top of a good hill, park my car at the bottom, ride a bicycle up the hill, and then cruise down it in my kart. A good workout, and a great reward for making it to the top!

START

PLANNING

It's OK to jump into this project with reckless abandon, but I recommend that you first make detailed drawings of your frame and suspension. Share your designs with people who might be able to guide you. When you've settled on a design, make cut lists for the tubing and other materials you'll need, and order at least 50% more than they indicate.

I used a Moleskine notebook and referred to the classic book, *How to Make Your Car Handle*, by Fred Puhn. Be careful when working out the geometry and placement of the suspension and steering systems; the front end can get very crowded with arms, shocks, tie rods, wheels, frame, etc.

Select wheels and tires based on the road your kart will run on. Tall wheels (like bicycle wheels) reduce friction by spinning more slowly than go-kart wheels, but they're also more likely to crumple under extreme cornering loads or bumps. Round tires are faster than flat (especially with high pressure) but more prone to slippage under tight cornering.

SEAT

I made the seat first, figuring it would give me something to sit in while I test-fitted everything around me. First make a paper mock-up. Lie on the ground on a large sheet of butcher (or similar) paper with your legs extended out in front of you. Trace where your body meets the floor from your upper back to your mid-thighs, and add perpendicular lines radiating away from it approximately 8"–12". Cut along these lines, then fold and tape the paper into a concave seat shape.

Transfer the design to thin sheet metal and cut it out with tinsnips. Bend the metal into shape and clamp overlapping sections together while you drill holes for the rivets.

Sheet metal edges tend to be sharp, so attach upholstery (and even padding) if you like. I secured black and red cloth-backed pleather around the seat's edges with riveted strips of sheet metal (Figure A), and added more rivets to hold the middle of the fabric against the seat. So far, my upholstered seat has held up to many months of sun, rain, and snow.

FRAME

The frame is welded together out of metal tubing, mostly ½" and ¾" EMT (electrical metallic tubing, aka thin-wall wiring conduit), but 1" steel pipe was used for the rear diagonal braces, lower side rails, and bottom crossbar, and 1¼" steel for the roll bar.

Cut the straight frame members with a chop saw. For ends that will be welded to other tubes, use a tubing notcher or milling machine to cut fish-mouths to fit (Figure B).

You can bend curved pieces made out of EMT or other lightweight tubing using a hand tube bender. On my frame, this included the rear harness cross-member, the 2 upper rails, and the 2 upper cross-members that hold the steering shaft. For the heavier roll bar, I used a hydraulic bender.

Weld the frame together. With EMT, sand off the galvanized surface layer where you're going to weld, as it can weaken the joint. Where suspension components, brake plates, and other parts attach to the frame, weld on tabs or tube stubs rather than cutting or drilling the tubes, which will make it weaker.

Give it a test fit (Figure C). Be sure to make car engine noises to pretend you're going fast!

SUSPENSION

For both front and back suspensions I used an unequal length A-arm design, which means that 2 pivoting horizontal arms connect each wheel to the frame, with the upper arm shorter than the lower one. This gives the wheels negative camber at bump and positive camber at droop, maximizing their contact with the road while cornering.

The shock absorbers bypass the upper arms and only attach to the lower arms, which need to be beasts because they bear the load of the entire kart. In back the shocks connect up to the sides of the roll bar, and in front they angle in and connect to a bracket sticking up from the middle of a horizontal frame member (Figures D and H).

The suspension arm pairs are mounted to the frame via ⅜" rods that rotate inside horizontal bushings welded to the frame. Both rods and bushings are made of higher quality steel so the arms swing up and down precisely, without moving side to side. At the wheel end, the arms connect to ball joints attached to the top and bottom of the kingpin (Figure E),

G

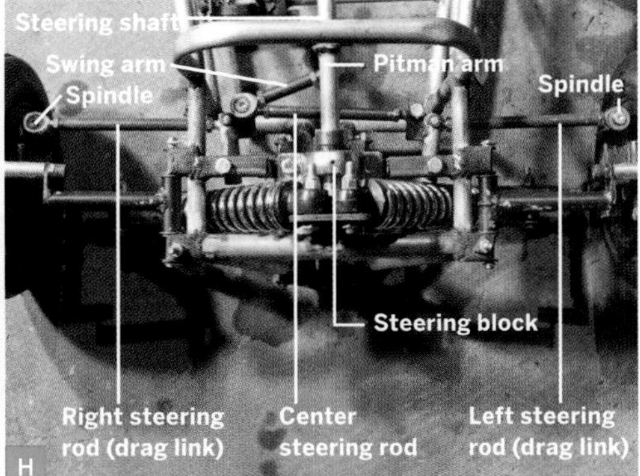

H

the vertical element that is the wheel's axis of rotation when the kart is turning.

To accentuate the camber change of the unequal A-arm suspension, I made kingpins slightly longer than the height difference between the upper and lower suspension mount points on the frame.

The kingpin is part of the spindle, a bracket that holds the axle and other components. Both kingpin and axle must be made of material sturdy enough to support the kart under stress. I used steel tubing.

STEERING

For small karts like this, a quick-release steering wheel hub makes it easier to get in and out of. Press the button, and the wheel pops right off.

The steering shaft must be held on in 2 places. I used pillow block bearings bolted to tabs welded onto bent cross-members on the frame (Figure F). The steering shaft disassembles into 2 pieces for easy removal.

At the steering shaft's forward end, a Pitman arm angles down perpendicular to the shaft (Figure G). The arm is a short, flat metal bar with one or more holes (I used 3) for bolting to the rest of the steering system through an adjustable tie rod. Attaching the swing arm to the inner hole on the Pitman arm makes the steering system the least sensitive, while bolting it to the outer hole makes it the most sensitive.

The swing arm runs to the right, where its other end attaches in a stack to 2 more tie rods: the right steering rod (aka steering arm or drag link), which connects to the right front wheel spindle; and the center tie-arm, which connects back across to the left steering rod, and by extension, the left front spindle (Figure H).

To make the steering and wheel alignment adjustable, I tapped all 4 tie rods with right-handed threading on one end and left-handed threading on the other. This lets you lengthen or shorten each connection length without removing the rod from the kart. The rear wheels also have tie rods, but these are just for alignment since they don't steer.

The steering rods connect to the spindle a few inches behind the kingpin so that while the kart turns, all 4 wheels angle around the same point. This geometry, known as Ackerman steering, minimizes tire slippage and skidding around curves. (With proper Ackerman geometry, the steering connection would be angled in rather than directly behind the kingpin, but my geometry approximates this for lower wheel angles, which are the ones you take at high speeds.)

SPINDLES

The spindles bring your suspension, steering arms, wheels, and brakes together. They represent the most precise part of the construction, and are especially involved if you want to minimize weight like I did.

Since I'm using drum brakes on the rear and no brakes on the front, my rear spindles connect the kingpin, axle, drag link arm

(wheel alignment), and brake mount. The front spindles carry the kingpin, axle, and steering rods.

To avoid sudden steering changes from going over bumps ("bump steer"), the steering rods must mount to the spindle so that they run perfectly horizontal — more specifically, parallel to and approximately the same length as the suspension arms when viewed from the front/back. Reinforce them well — I had problems with steering rods twisting on big bumps until I welded on some gussets that double as backing plates for mounting the brakes.

Use your most precise tools when drilling holes for spindle parts; use jigs and measure the angles of everything when you weld them together. If the components aren't all at their correct relative angles and positions, the steering and suspension will suffer.

BRAKES

The brake pedal connects to the center of a pivoting balance bar that pulls the brake cables from each end. This ensures that both sides of the kart get equal braking force. A perforated tab on the rear spindle holds the brake end of the cable housing (Figure I).

Brake cable housing dissipates energy through friction and compression, especially around bends. To maximize braking power and responsiveness, you should route cable as straight and with as little cable guide as possible, even though using one long run of cable housing is easier.

SEAT AND BELTS

Pick at least 4 places (preferably more) to mount your seat to, and attach it using screws rather than rivets. If you ever have to change something, you'll be glad you did. Weld on mounting stubs rather than drilling and tapping into the frame. Drill a drain hole in the bottom of the seat, especially if you plan on leaving it outside.

Seat belts are the most important part of the kart. A five-point racing harness — side belts, shoulder belts, and crotch belt — will keep you secure relative to the kart whether

I

⚠ WARNING:
Wear the proper safety gear when driving a kart like this, which can reach highway speeds and doesn't have as much crumple room as a car. A motorcycle helmet is an absolute must, and I highly recommend gloves, a neck brace, closed shoes, and tough clothing.

you're on the road or in a crash.

Attach belts to the strong parts of the frame. I bolted the side and crotch belts to steel frame tube, then welded the bolt in place. My shoulder belts strap onto pipe rather than threading through a bracket, so I welded on metal loops that prevent them from sliding sideways off my shoulders.

UNDERBODY

I added an underbody to keep debris from flying up and prevent my feet from striking the ground. I simply laid a piece of sheet metal underneath the frame and bent it up on either side. This one-piece construction means there are no seams to impede airflow or collect dirt. It also looks nicer, and you can get by with fewer attachment points than with more complicated underbodies.

FINISHED!

Now drive off into the sunset — downhill, of course — and make sure you have cameras recording. I use GoPro Hero HD cameras to film and a Garmin Edge 305 cycling computer to log my speed and location.

+ See makezine.com/go/warrrior for complete build instructions, including tools and materials lists, and visit vimeo.com/user1346886 for videos of the Weekend Warrior in action.

Jeremy Ashinghurst of Towson, Md., is an electrical engineer for DeWalt who spends his free time driving karts, dabbling in photography, and hanging out at Harford Hackerspace.

1

2

3

4

MAKE: PROJECTS KARTS & WHEELS RUNNERS-UP

We had a lot of fun with our Karts and Wheels contest. Check out Jeremy Ashinghurst's winning project on page 60, and peep the runners-up here. Plus, keep your ears to the wire for our new robot contest at makeprojects.com!

1. The Bike Buh Cue
By Tho X. Bui makezine.com/go/bikebuhcue

2. Quad Roller Skates
By Brad Wong makezine.com/go/quadskates

3. Teardrop Camper Trailer
By Werner Strama makezine.com/go/camper

4. Soapbox Racer from Found Material
(Honorable Mention) By Monika Wuhrer
makezine.com/go/foundsoapbox

MADAGASCAR INSTITUTE CHARIOT RACES

BY NICK NORMAL

Complete with flag-bearers, arena emcees, and sounds of marching bands, these games at World Maker Faire New York 2010 were reminiscent of their Roman and Greek predecessors but without the equines. Instead there were motorized carts, pedal-powered jousting bikes, human wheelbarrows, ball-and-chain-wielding gladiators atop a motorcycle-driven fish boat (on land!), shopping cart mods, and a giant kraken-inspired chariot with many tentacles that was very much on the offensive. The motto was: "You're not gonna win with brakes!" madagascarinstitute.com

Photography by Scott Beale (chariot races)

UNAUTHORIZED FUN ON WHEELS

BY TODD LAPPIN

In the realm of simple machines, an illegal soapbox derby car is about as simple as it gets: just combine an inclined plane with four wheels, and you're off to the races. Beyond that, anything goes and everything is optional — except for a set of brakes, a helmet, and (depending on your jurisdiction) a beer-can holder.

The cars themselves vary considerably. Some are aerodynamic speedsters with welded-steel frames optimized to minimize friction and maximize momentum. Some are ruggedized war-wagons built to withstand the rigors of full-contact, wheel-to-wheel competition.

Others are more kinetic sculpture than race car, built to make a gravity-powered artistic statement while rolling awkwardly downhill.

Illegal soapbox derby racing isn't so much "illegal" as "unauthorized." Races are held on public roads in early-morning hours to minimize the risk of chance encounters with hostile neighbors and oncoming traffic.

In contrast to the rigidly run races organized for kids by uptight sanctioning bodies like the All-American Soap Box Derby, the illegal variant is a more informal and aggressive affair. Regulations vary, but the rulebooks for illegal soapbox derby races — if they exist at all — are designed primarily to satisfy adults who don't much care for following the rules. Whining is always frowned upon.

Many trace the roots of illegal soapbox derby to San Francisco, where pranksters began races in the 1970s.

Races still take place in the Bay Area, but the center of gravity for illegal soapbox derby racing has arguably moved to Southern California, where the San Fernando Valley Illegal Soap Box Federation (sfvisbf.com) organizes rough-and-tumble competitions on the region's winding mountain roads.

But really, soapbox racing can happen almost anywhere. All it takes is a big hill, some mechanical ingenuity, and a gravity-fueled need for speed.

GASIFIER GO-KART

BY KIPP BRADFORD

Great engineering design work is being done by students, including a gasifier-powered go-kart project undertaken by Brian Fisher, David Gagnon, and Devin Sutcliffe from Brown University. The project goal was a one-person vehicle that demonstrates the use of alternative fuels in an internal-combustion engine.

Wood chips contain plenty of energy, but that energy must be converted from solid, fibrous wood into a form that can flow through a car's fuel lines. The process is called *gasification*, where wood is heated until

it partially combusts, causing it to release hydrogen and carbon monoxide gases.

After fabricating the steel combustion chamber, mounting it on a surplus go-kart, and adding a few tweaks to filter out smoke, the Brown students successfully drove the go-kart on wood chips. Gaseous! makezine.com/go/gasifier

Photography by Todd Lappin (soapbox)

WHEELCHAIR ACCESSIBLE
BY GOLI MOHAMMADI

When Joel Sprayberry's beloved dachshund, JubalLee, hurt his back, Sprayberry decided to make his canine companion a wheelchair (well, more like a wheeled cart) to help him get around until he regained the use of his back legs. For about $20 in supplies — including tent poles, tennis racket padding, pneumatic wheels, and a dog harness — you can follow Sprayberry's how-to, and your pup can be speeding along in no time.
instructables.com/id/dachshund-wheelchair

THE BRILLIANT MOONBEAM
BY LAURA COCHRANE

Retired boat captain Jory Squibb of Camden, Maine, built a micro-car that gets 85–105 miles per gallon.

The 69-year-old put it together in his garage during the winter of 2005–2006, using two old Honda scooters. Weighing 400lbs, the 79"-long, 52"-wide, 56"-high vehicle, named *Moonbeam*, has a 2-gallon gas tank capacity and can zoom along at a respectable 25mph–40mph, with a maximum speed of 53mph on level ground.

It's been driven more than 13,000 miles, and Squibb remarks that it's used by his family almost every day, especially by Chloe, his youngest daughter, "who finds it a great guy magnet."
makezine.com/go/moonbeam

REDEFINING STREET LEGAL
BY GOLI MOHAMMADI

Inspired by a luge he saw on TV, Canadian Gerry Lauzon decided to relive his old skateboarder hill-bombing days and build one. Older and wiser, he wanted to add brakes and make sure it was street legal.

In Canada, street luges are considered skateboards, which are illegal on public roads, unlike bikes. The legal definition of a bike specifies having at least one brake and a bunch of reflectors.

He fashioned his luge in an afternoon using a trash bike, some 2x4s, plywood, and rollerblade wheels. He even used the serial number on the donor bike's bottom bracket, and registered it as "Lauzon Cycle, Model 008" (his eighth build). Lauzon hit his goal of breaking the 60kph barrier with his son and a friend.

"Kids need danger in their life or they'll always be afraid of everything," Lauzon says.

Check out his build notes, and make some street legal danger of your own.
makezine.com/go/luge

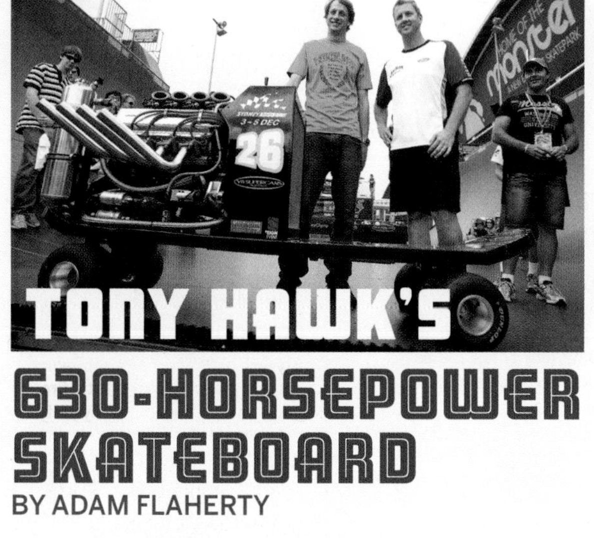

TONY HAWK'S 630-HORSEPOWER SKATEBOARD

BY ADAM FLAHERTY

The full-throttled madmen from V8 Supercars of Sydney, Australia, fabricated their monster 630-horsepower V8 Skateboard for a recent visit to the Telstra 500 by the legendary Tony Hawk. Never deterred, Mr. Hawk literally jumped onboard amidst a crowd of onlookers. Fortunately, for safety reasons, the beast is speed-limited. makezine.com/go/v8skate

HOME-BREWED SEGWAY SCOOTER

BY JOHN BAICHTAL

Charles Guan, a mechanical engineering senior at MIT's Media Lab, built the Segfault, a self-balancing electric vehicle with a water-jetted aluminum chassis.

It packs 9" scooter wheels, a pair of 27:1 gearmotors, and a 24V LFP battery pack. The Segfault balances thanks to a custom-designed filter that combines data from an accelerometer, whose sensitive axis lies perpendicular to the wheel axles, and a gyroscope, whose axis of rotation lies parallel.

As the Segfault rolls forward, responding to the rider's lean, the gyroscope helps prevent the accelerometer from overcorrecting. Next up, Guan hopes to cram a bigger battery pack into the chassis. makezine.com/go/segfault

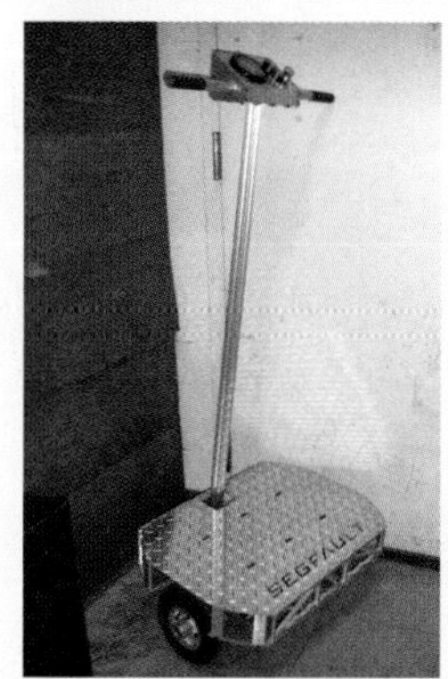

GO, TOY RACER, GO!

BY GARETH BRANWYN

At last year's Maker Faire Detroit, one of the more entertaining and madcap events was the Power Racing Series, organized by Chicago hackerspace Pumping Station: One. This second-annual race challenged all hackerspaces and maker groups to modify Fisher-Price Power Wheels or Little Tikes Ride On vehicles to make them adult-bearing and race-worthy.

To be a participating team, you had to have at least three members. Teams had a budget limit of $500 (including the base vehicle) and had to keep the wheels, hubs, and tires stock. Body modifications (within size and safety limits) were encouraged, and electric motors and batteries (sealed, maintenance-free batteries only) could be upgraded.

In 2010, eight teams competed in four events, with hackerspace i3 Detroit taking home the coveted trophy.

The race is on again for Maker Faire Detroit 2011. Event organizer Jim Burker told MAKE that, at press time, there were already seven teams signed up to compete.

So zip up those mechanic coveralls, grab a wrench, practice your smack-talkin', and stuff your big adult butt into a kiddie car. You can find info about last year's event and download the hysterical "Electrical Fires and You" Power Racing Series Guidebook from powerracingseries.org.

Photography by Michelle Andonian (Power Racing Series); Getty (Tony Hawk)

ROBO-RAINBOW

BY NICK RAYMOND

Robo-Rainbow is an "instrument of mass destruction" designed by Alexander Kurlandsky to deliver picture-perfect rainbows all over town. Connected to a bike, the metal frame supports a telescoping arm with six cans of spray paint that are counterbalanced by bricks. The arm pivots 180°, driven by a cordless drill, gears, and chain, to trace an arch. An R/C transmitter sends a signal to an Arduino on the cart and controls the servos mounted to the spray-paint caps. Two limit switches input a signal to the Arduino when the arm reaches the end of the rainbow, killing power to the drill. Mechanics, electronics, and rainbows — beautiful. vimeo.com/19374769

DIGITAL-TO-ANALOG CAR MOD

BY GARETH BRANWYN

Racer 0.2 is an analog racing game based on the classic *WipEout*, but players control the physical world from within a virtual one. A modular racetrack is built almost entirely of cardboard, and camera-equipped, radio-controlled cars zip around it, controlled by a driver watching video inside of an arcade racing console. racer.sputnic.tv

BIOFUEL ELECTRIC SNOW HUMMER

BY KEITH HAMMOND

How do you drive across 1,000 miles of ice without fossil fuel? Ask Nick Baggarly. His nonprofit group Drive Around the World organizes expeditions for charity, and they're headed for the South Pole on biopower.

After circling the globe in Land Rovers, Baggarly wanted to quit petroleum. "I've been in thousands of cities and traffic jams all over the world," he says. "It gets you thinking about all those millions of little fires burning on the road. It's really just not sustainable."

The Zero South Polar Traverse Vehicle is a hybrid Hummer H1 on snow tracks, with a 150kW generator spun by a Steyr turbo diesel burning biofuel. To build it, Baggarly assembled veterans from General Motors, AC Propulsion, Jay Leno's Garage, and K&N Engineering. They tossed the stock engine and tranny and flipped the differentials to make room for two UQM electric traction motors and 16 ElectroVia 48V lithium battery modules.

The build is open source, so you'll be able to make your own super hybrid. "We want to inspire young people," explains Baggarly. "They're going to be doing the car restorations, dropping in Hemis or Duramaxes — we want to encourage them, what if we used electric motors, lithium batteries? That's how this is going to happen." makezine.com/go/hummer

» See the Zero South Hummer at Maker Faire Bay Area, May 21–22: makerfaire.com

1+2+3 Bokeh Photography Effect

By Sindri Diego

You can make it!

BOKEH COMES FROM THE JAPANESE word for "blur." In photography, the bokeh effect has to do with the aesthetics of out-of-focus areas of the picture.

With this project, you can create an effect that makes the out-of-focus lights in your pictures appear any shape you want.

YOU WILL NEED

- Paper, thick and black
- Camera lens, 50mm
- Camera
- Compass
- Ruler
- Scissors
- X-Acto knife

1. Measure and cut a paper disk.

Set your compass to measure 25mm between the spike and the pencil, and draw a circle 50mm in diameter.

Cut out the circle, leaving a little tab of paper somewhere on the edge to use as a handle. If you use the paper wisely you can make many bokeh disks from one sheet.

2. Cut your shape out of the center.

Draw your shape, centered, on the 50mm paper disk. It's best if the shape is not too complicated.

Using the X-Acto knife, cut out the shape you drew. Be careful to not cut yourself.

3. Start making cool bokeh effects.

Position your bokeh disk in front of the lens. You can probably fix it in place using the threading that's intended for attaching filters. It doesn't matter how close the paper is to the glass lens, but it has to cover the whole lens.

Adjust your camera to the lowest aperture setting and start shooting. Remember, you want to have the lights out of focus to get the effect. Be as creative as you can — if you have a digital camera, you can always delete the images you don't like.

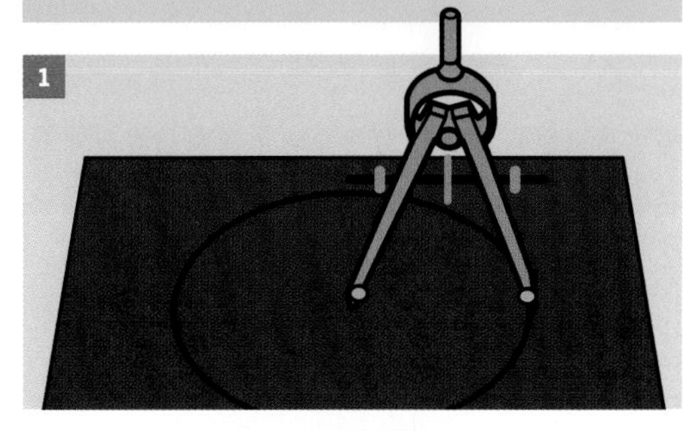

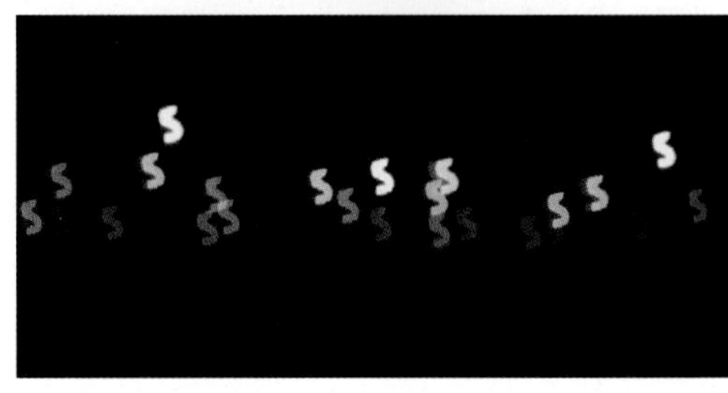

More bokeh effect photos: makezine.com/26/123_bokeh

Build and discuss this project: makeprojects.com/project/bokeh/371

Sindri Diego is a 19-year-old Icelandic multimedia design student, gymnastics champion, and coach. He is taking his first steps in photography and loves to try new things with it.

Illustration by Damien Scogin; photograph by Sindri Diego

Make: Projects

Make a Rubens tube and watch its mesmerizing flames dance to the music as they heat up the night. Then solder up a pocket-sized noise looper that twists and layers fascinating rhythms with two knobs and a button. Need to refuel? Turn a sunny window into a continuously productive spirulina superfood farm.

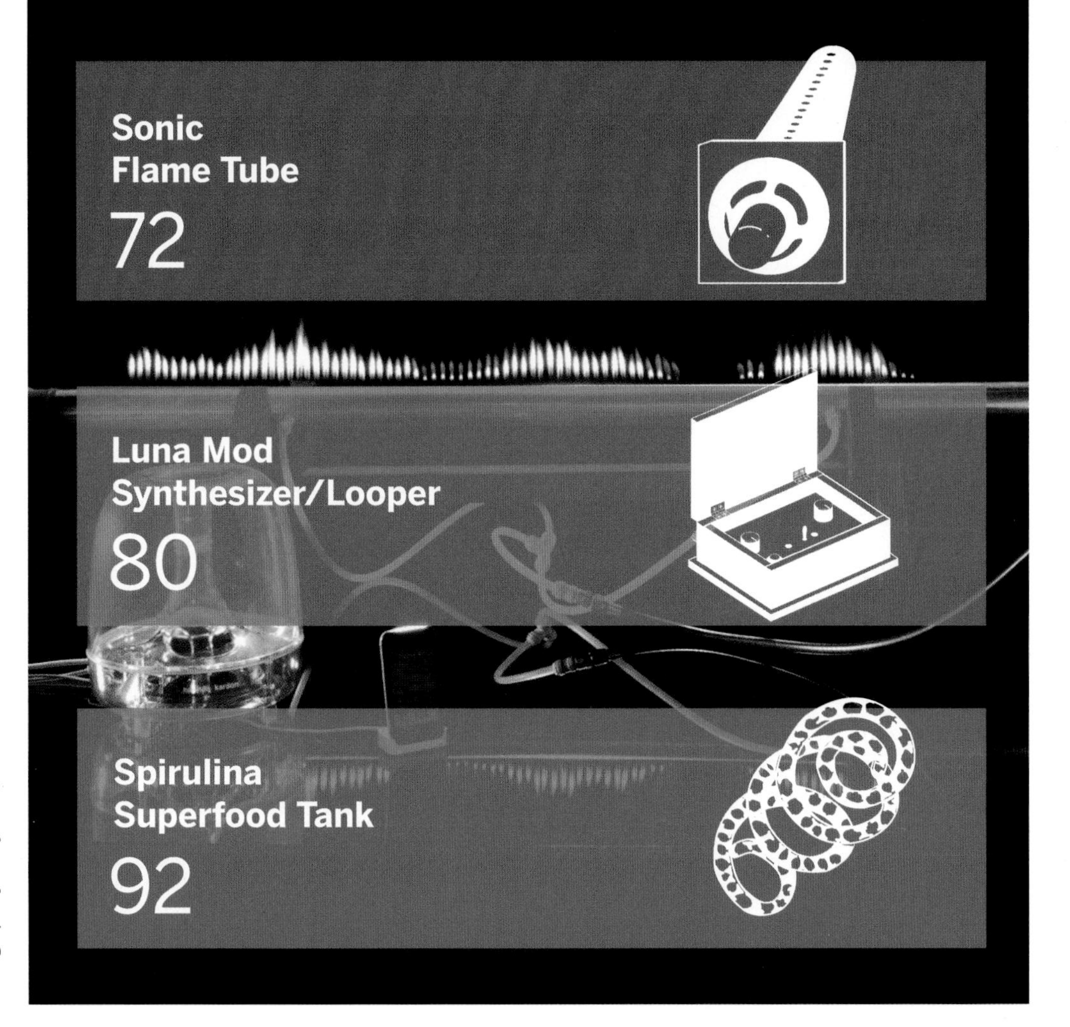

Sonic Flame Tube 72

Luna Mod Synthesizer/Looper 80

Spirulina Superfood Tank 92

Photograph by Garry McLeod

THE FLAME TUBE

Visualize Waveforms with Fire

By William Gurstelle

Fiery devices have always fascinated me. From Jam Jar Jets (MAKE Volume 05) to Fire Pistons (Volume 19) to Faux Flames (Volume 25), I've built all sorts of fire-related projects. So when a friend told me about a device that lets you visualize sounds using fire, I knew I had to make one. I found it described in several old physics demonstration manuals, then I adapted those directions to make it less expensive and easier to build.

When you play a constant-frequency tone into the Flame Tube, it displays a perfect sine wave of fire. Play music, and the flames make a wild display caused by big, air-moving bass beats, standing waves from resonant frequencies, and other acoustic phenomena. It's inspiring, fun to watch, and good for heating up your garage or workshop on a cold day.

In 1860, Dutch physics professor Pieter Rijke was investigating the relationship between sounds, gases, and fire. He stuffed a piece of iron mesh inside a large glass tube, then held it over a gas flame until the mesh was red-hot. Suddenly, the contraption emitted a sustained musical tone so loud that workers several rooms away complained.

Intrigued, Rijke's colleagues set out to discover the reason for "the singing flames." Some thought it was the periodic evaporation and condensation of water, but later scientists showed that the sound was caused by waves of air, set in recurring motion by the fire's heat. Hot air, being less dense, moved upward while cool air sank. This vibrating air resonated at the natural frequency of the tube.

Years after Rijke's work, German scientist Heinrich Rubens turned the idea on its head. He knew fire could produce resonating sound waves. Was it was possible to use fire to make sound waves visible for the first time? In his laboratory at the University of Berlin he developed the standing wave flame tube — also called the Rubens tube in his honor.

William Gurstelle is a contributing editor to MAKE. His new book, *The Practical Pyromaniac*, is available beginning in early June. thepracticalpyromaniac.com

SET UP: p.75 | **MAKE IT: p.76** | **USE IT: p.79**

Photograph by Garry McLeod

THE SHAPE OF SOUND

Turn sound waves into standing pressure waves — expressed in fire.

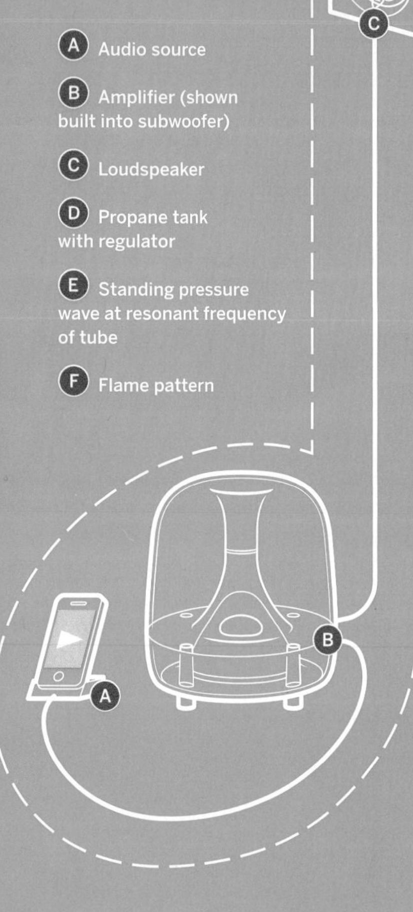

The flame tube (aka Rubens tube) is a waveform visualizer. It works because sound is a pressure wave. As sound moves through a gas like propane, the wave alternately compresses and expands the gas in different regions. When you use a frequency generator to produce a constant tone of, say, 440Hz (the musical note A), the speaker pushes this sound through the gas, and a stationary wave is set up.

The stationary wave causes areas of high pressure to appear at fixed points along the pipe, spaced half-wavelengths apart. Where the pressure is high, the propane is driven more strongly out of the pipe, resulting in a tall flame. Between these high-pressure points will be low-pressure points that create lower flames.

When music is played instead of the frequency generator output, the clean curves of the sine wave are replaced by much more chaotic and perhaps even more intriguing patterns. Strong vibration from drums and low-frequency sounds from bass guitars, tubas, and string basses cause the flame tube to send fire pulses out of the holes nearest the loudspeaker. Overlaying that are sine waves that are visible whenever the musical pitches being played coincide with the tube's resonant frequencies. This layering of resonant frequencies and bass beats produces a dazzling display of musical pyrotechnics.

Illustration by James Provost

SET UP.

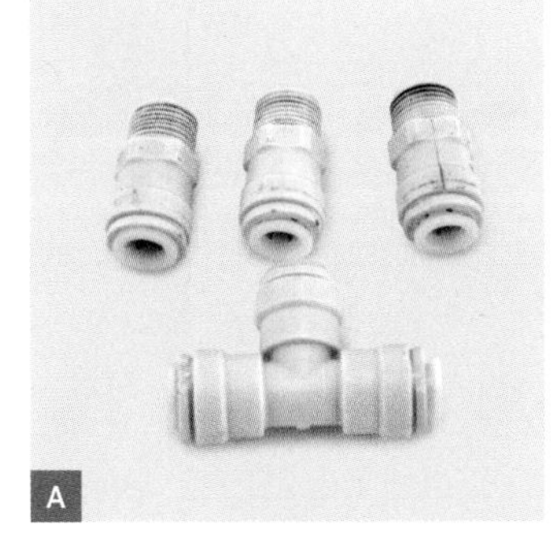
A

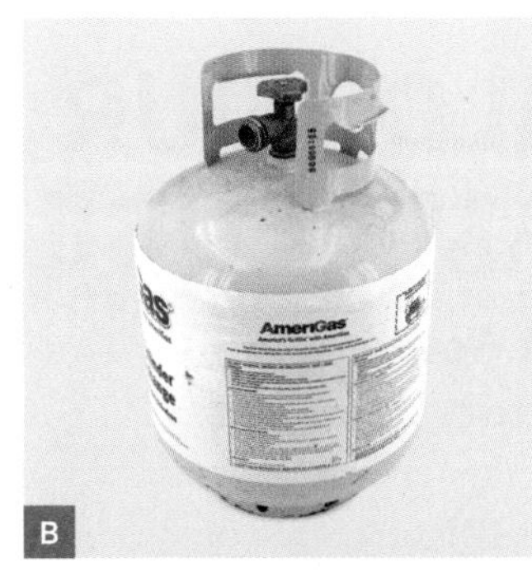
B

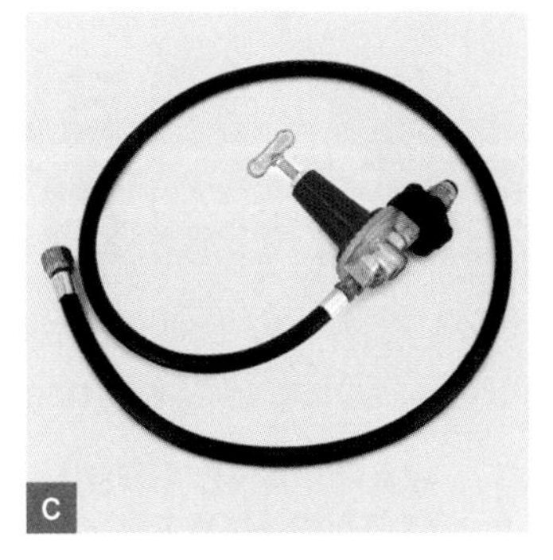
C

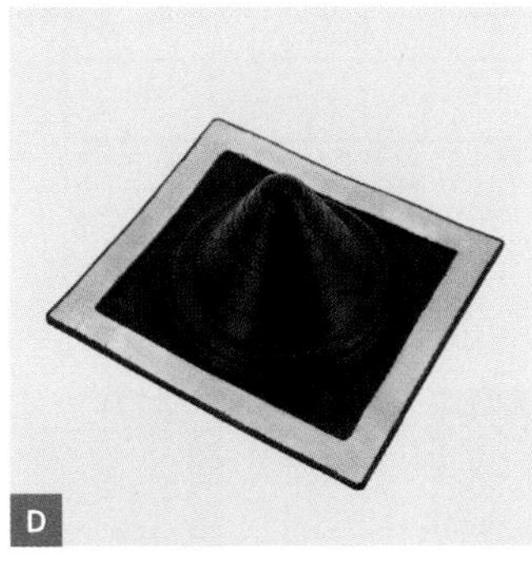
D

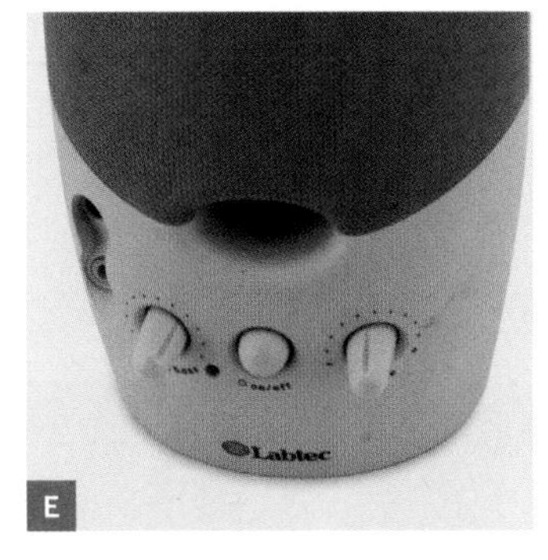
E

F

MATERIALS

A. Plastic tubing fittings, ⅜" OD, push-to-connect (PTC): tee connector (1), adapters to ⅜" male NPT pipe thread (2), adapter to ⅜" flare (1) These are found in the plumbing aisle at hardware and home stores.

If you can't find the ⅜" OD PTC × ⅜" flare adapter, you can buy a third ⅜" OD PTC × ⅜" male NPT adapter and use that to connect a brass ⅜" female NPT × ⅜" flare adapter, as we'll show here. Or improvise some other combination of PTC to NPT to flare fittings.

B. Propane tank, 20lb

C. Gas regulator High-pressure gas regulators such as those used in high-output outdoor stoves work well if throttled down. However, the propane regulator used on a standard gas grill will work. Note that there's a safety device in the regulator that shuts off the gas if you open the valve too quickly. To prevent this, open the valve very slowly.

D. Pipe flashing boot (optional) Sometimes called a "witch hat," this conically shaped piece of rubber is designed to fit pipes of differing diameter together, so it's an easy and secure way of attaching a 2" conduit or pipe to a larger-diameter speaker. They're available at home stores in the roofing materials aisle.

E. & F. Amplifier and loudspeaker A small, monaural amp and a 3" speaker are plenty. You could salvage an amplified computer speaker (shown here), or feel free to use larger old hi-fi equipment.

NOT SHOWN

- **Plastic tubing, ⅜" OD (outer diameter), 1 roll**
- **Steel conduit, 2" diameter, 5' length** You can use steel pipe instead, but don't use plastic pipe — the heat from the flame tube will soften and melt it.
- **Rubber balloons, helium quality** Helium-quality balloons are thicker and less likely to leak than regular balloons.
- **Rubber bands and/or strong tape**
- **Dimensional lumber, 2×4, 78" total length** cut to lengths of 12" (4) and 30" (1)
- **Deck screws, 2½" (1 box)**
- **Frequency generator and/or music sources** Pure-frequency audio test tones can be found online as *.mp3* or *.wav* files. Free or inexpensive frequency generator applications are available for personal computers, iPhones, iPads, and other handheld computing devices. Any music source you can connect to the amp will work.

TOOLS

- **Electric drill or drill press** While a handheld drill will work, 100 is a lot of holes to drill!
- **Drill bits: 0", 9"**
- **Driver bit** to fit the deck screws
- **Pipe tap, 2-18 NPT** This is a pipe tap. It's not a 2" hole tap. The correct tap makes a hole that fits the exterior of a 2" pipe.) You'll also need a handle to turn the tap.
- **Tape measure** for laying out the holes
- **Pipe thread compound**
- **Marking pen**
- **Safety glasses**
- **Lighter, long-handled**

Photography by Sam Murphy

MAKE IT.

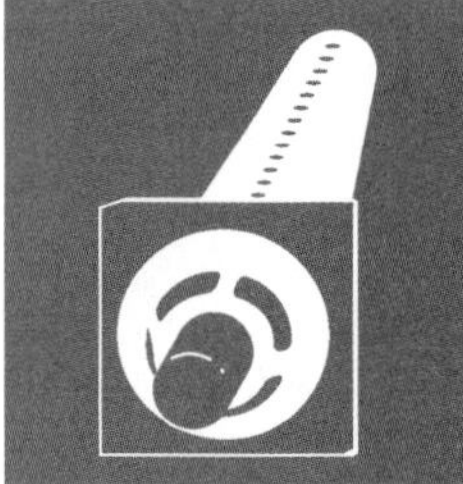

BUILD YOUR FLAME TUBE

Time: A Weekend
Complexity: Moderate

1. MARK AND DRILL FLAME HOLES

1a. Beginning 6" from the end of the conduit, make a series of marks with the marking pen, ½" apart in a straight line extending across the top of the conduit. Stop marking 6" from the other end.

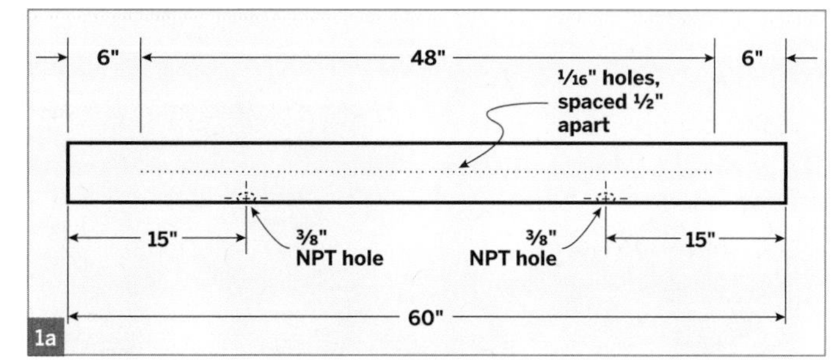

1a

1b. Drill 1/16" holes on the marks, taking care to make the holes as perpendicular to the surface of the conduit as possible (aim toward the center of the conduit). This will take a while, as there are nearly 100 holes to drill. To reduce drill wobble, move the drill bit up into the chuck to shorten the exposed length of the bit.

1b

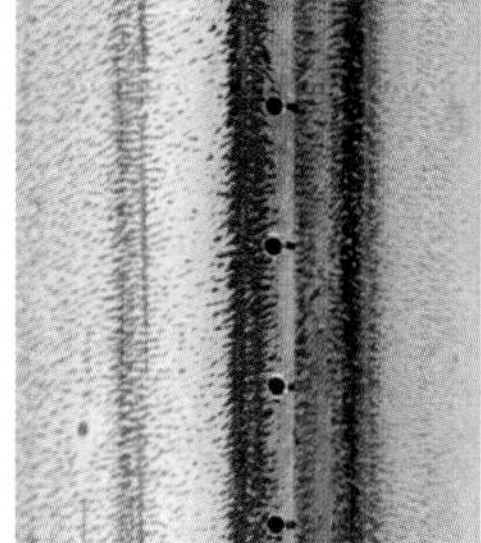

2. DRILL AND TAP STEEL CONDUIT

2a. Rotate the conduit 120° and drill two 9/16" holes, 15" from each end. Again, take care to make the holes as perpendicular to the surface of the conduit as possible. Tap the holes with the 3/8-18 NPT tap.

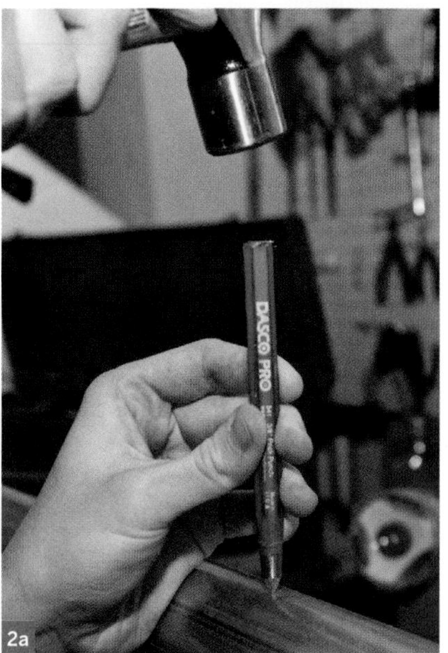

2a

TIP: To tap a hole in metal, place a drop of oil on the threads of the tap and then insert the tap into the hole.

2b. Turn the pipe tap clockwise ¾ turn, then turn it counterclockwise ½ turn. Continue this process until you're able to thread the entire hole with the pipe tap.

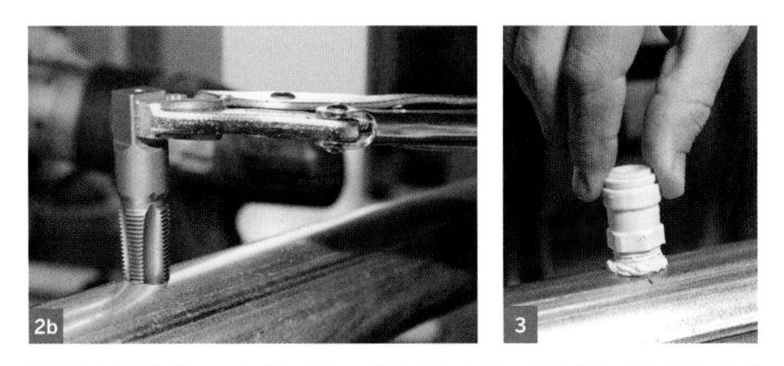
2b
3

3. ATTACH THE GAS FITTINGS

Smear pipe thread compound on the threads of the ⅜" push-to-connect (PTC) to ⅜" male pipe thread adapter fittings, and screw them into the 2 tapped holes until they bottom.

4. PLUMB THE GAS SUPPLY

4a. Cut two 30" lengths of plastic tubing and insert one into the PTC side of each plastic fitting on the conduit. Insert their opposite ends into a PTC tee fitting. Insert a 12" length of plastic tubing into the remaining open port of the tee fitting.

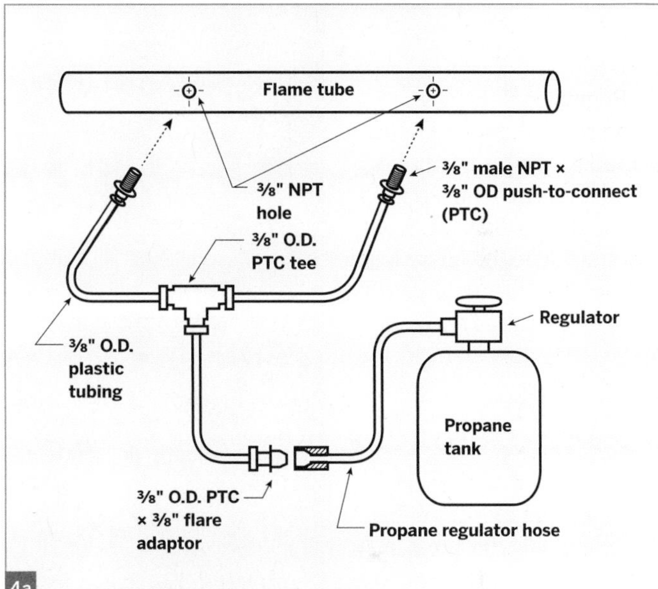

4a

4b. Insert the other end of the 12" tube into the PTC side of the flare fitting (or your own combination of fittings that begins with PTC and ends in the ⅜" flare).

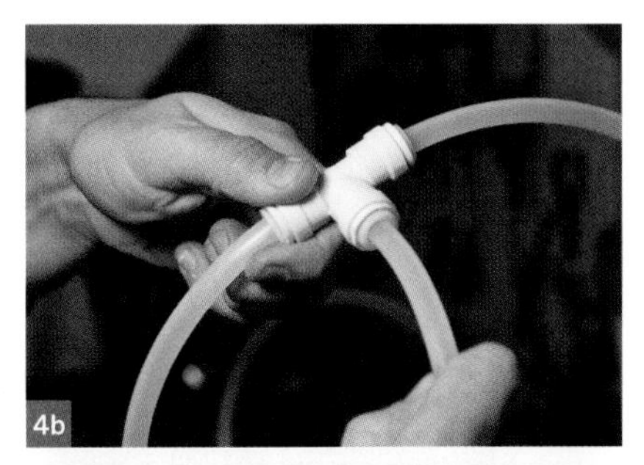
4b

5. ATTACH THE BALLOONS

Cut off the narrow end of 2 balloons and discard. Attach the remaining pieces of the balloons to each end of the conduit. Stretch them tight, and secure each with a rubber band or tape. We used red electrical tape here, for looks.

5

6. BUILD THE WOODEN STAND

6a. Begin by placing 2 of the 12" wood pieces into an X shape.

Secure with two 2½" deck screws. Repeat to make a second X with the remaining pieces. Connect the X brackets to one another with the remaining 2×4, using the deck screws.

6a

6b. Place the conduit on the stand with the holes pointing up. Make certain the area around the conduit is at least 5' — that's 5 feet — away from any combustible materials on all sides, as well as above it.

7a

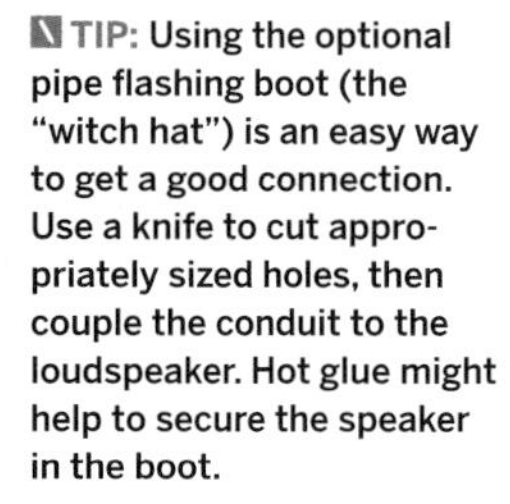

TIP: Using the optional pipe flashing boot (the "witch hat") is an easy way to get a good connection. Use a knife to cut appropriately sized holes, then couple the conduit to the loudspeaker. Hot glue might help to secure the speaker in the boot.

7. CONNECT SPEAKER TO TUBE

7a. Place the loudspeaker firmly against one of the balloons and affix the speaker to the end of the conduit. The quality of this connection has a large impact on how well the Rubens tube works; you may need to use duct tape or hot glue, or form a flexible rubber gasket to affix the speaker securely.

7b

7b. Wire the speaker to the amplifier and frequency generator.

8. CONNECT THE PROPANE

Attach the flare fitting to the propane regulator and propane tank. The flame tube is now ready to use!

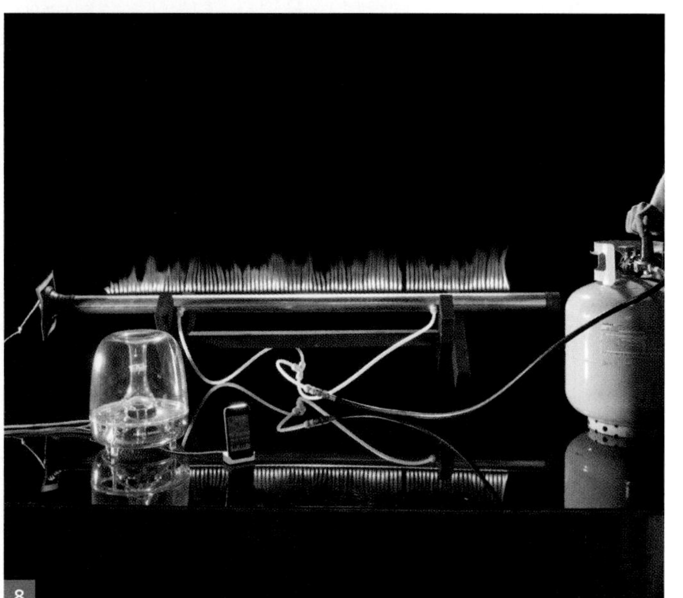

8

Photography by Garry McLeod (Figure 8 and opposite)

USE IT.

FIRE UP THE TUNES

Testing

Open the gas valve, slowly. Using the long-handled lighter, ignite the gas jetting out of the exhaust ports (the holes at the top of the tube). Adjust the gas pressure so each flame is about ¾" high. The height of all flames should be equal. If they're unequal, turn off the gas and clean any clogged holes by reaming them with music wire or redrilling.

Turn on the frequency generator and amplifier and set it to 440Hz. The frequency of the sound wave should be visible as a pattern of repeating high and low flames jetting from the holes on the pipe.

Experiment with different frequencies to see different patterns and wave shapes.

⚠ SAFETY NOTES

1. This project involves open flame and uses propane. It should be attempted only by adults, or under the close supervision of adults. The metal surfaces get very hot; do not touch them until they are cool.
2. Operate this device on a nonflammable surface and keep combustible objects well away from the flames.
3. Use the flame tube only in a well-ventilated space. Gas leaks are a possibility.
4. Keep a fire extinguisher handy.

Play Some Music

Replace the frequency generator with a music source and experiment with different types of music to understand how they affect the pulse and shape of the flames.

If you have trouble keeping all exhaust ports lit during loud musical passages, try covering the last 6" of holes on both ends with aluminum tape.

See a video of the flame tube in action: makezine.com/26/flametube

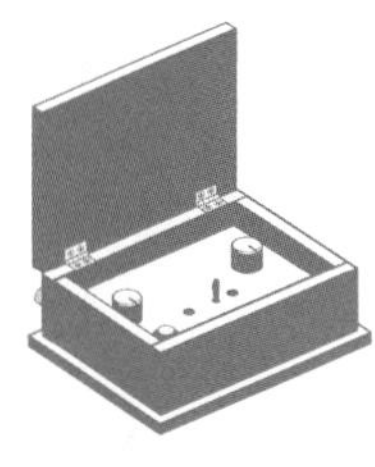

THE LUNA MOD

Simple, Addictive Sound Effects Looper

By Brian McNamara

This simple handheld synthesizer and looper box generates intriguing sonic rhythms, and you'll be amazed at how easy it is to play.

The Luna Mod is an easy and fun instrument that will have you making great-sounding loops in no time. Rather than sampling input like a traditional loop station, the Luna Mod synthesizes its own sounds, and you play it using two knobs and one button.

I based the Luna Mod on the Wicks Looper, which I sell on Etsy, but it's designed to be even simpler to build and play, without any complicated functions that would never get used. One knob controls the sound generated, one knob controls the tempo of the loop, and the button writes the current sound into the ongoing loop. The variety of sounds you can get from these three controls is amazing.

You don't need any special or expensive tools to build the Luna Mod, and as with all my other electronic instruments, I designed it for hackability.

At the heart of the Luna Mod is a $3 component — an 8-pin Picaxe 08M micro-controller. This handy device's built-in programming port lets you easily change the sounds it produces and also change which variables are tied to the controls.

Brian McNamara (@rarebeasts) runs a small electronics design lab in Canberra, Australia, called Rarebeasts (rarebeasts.com). He builds a range of handmade electronic devices, mostly musical instruments, many of which he plays in his noise band.

Photograph by Garry McLeod

SET UP: p.83 | MAKE IT: p.84 | USE IT: p.91

PLAYING THE 'AXE

Inside the Luna Mod, the Sound and Tempo pots connect to the Picaxe microcontroller's 2 analog-to-digital converter input pins. The Picaxe code reads these 2 values, along with the state of the Write pushbutton, to determine the sounds it produces.

- **A** 9-volt battery supplies power.
- **B** Voltage regulator converts 9V DC into the 5V DC needed by the circuitry.
- **C** Power switch turns device on and off.
- **D** Power LED shines when device is on.
- **E** Tempo knob sets the loop tempo.
- **F** Tempo LED blinks once per loop.
- **G** Sound knob sets the pitch and quality of the sound.
- **H** Write button plays sound (per current sound setting) and writes it into the loop when pressed.
- **I** Picaxe microprocessor runs Luna Mod software to read controls and generate the sound loop.
- **J** Programming port uploads Luna Mod software from the computer to the microprocessor.
- **K** Output jack lets Luna Mod plug into amplifier, etc.
- **L** Attractive wooden box holds Luna Mod workings while matching any decor.

Search YouTube for "Luna Mod" to see it work and hear audio samples.

Luna Mod Software

The Luna Mod's software, which is only about 30 lines of code, loops endlessly through Picaxe memory registers 80–127, each of which holds an integer between 0 and 255. The Tempo knob sets the wait time at the end of each loop.

When the Write button is down, the program stores the value read from the Sound knob into the current memory register. Then the program sends the current register value (whether or not it was just changed) to the Sound function, which controls the Picaxe's internal resonator. The value 0 (the default) produces no sound, 1–127 produce ascending tones, and

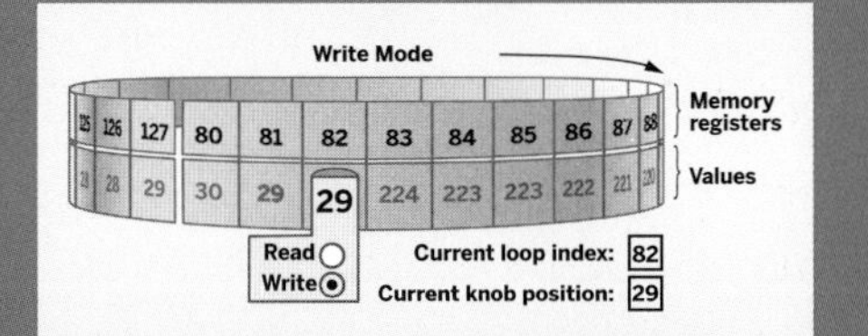

128–255 generate ascending white noises. The program then calls Sound again with the value plus 128, which sounds cooler; playing the basic tone and white noise together produces a full, less tinny sound.

Illustration by Timmy Kucynda

SET UP.

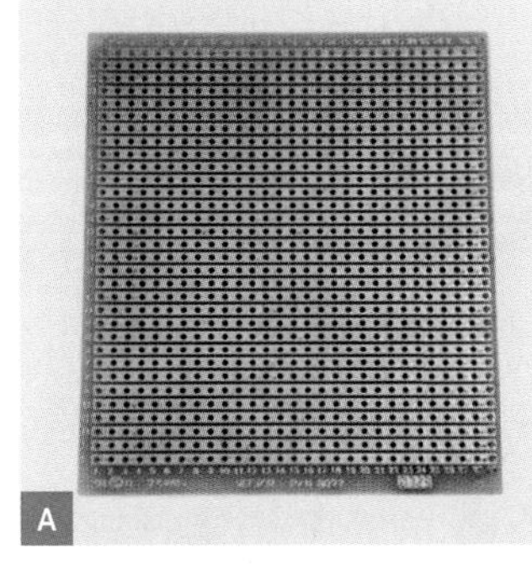

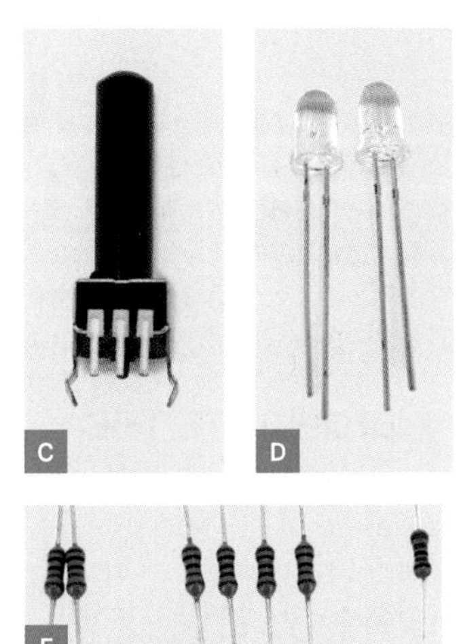

MATERIALS

A. Stripboard, 26×16 hole can be cut from a larger board, such as 3"×3.5" #V2018-ND from Digi-Key (digikey.com), or the 2"×5" 1000L Phenolic from Veroboard (veroboard.com)

B. Microprocessor, Picaxe 08M #COM-08308 from SparkFun Electronics (sparkfun.com)

C. Potentiometers, vertical PCB mount (2) part #COM-09288 from SparkFun

D. LEDs, 5mm, blue (2) #2006764 from Jameco Electronics (jameco.com)

E. Voltage regulator, LM78L05 part #51182 from Jameco

F. Resistors: 1kΩ (2), 10kΩ (4), 22kΩ (1)

G. Capacitors, ceramic: 10nF (1), 100nF (1) aka 0.01μF and 0.1μF

NOT SHOWN

- **Micro-mini toggle switch** #275-624 from RadioShack (radioshack.com)
- **IC socket, 8-pin**
- **Wire, 22 gauge, stranded and solid, multiple colors**
- **Stereo inline audio jack, 1/8"** #274-274 from RadioShack
- **Switch, momentary pushbutton** #275-644 from RadioShack
- **Audio jack, mono, 1/4" panel-mount** #274-252 from RadioShack (2-pack)
- **Battery clip for 9V battery**
- **Battery, 9V**
- **Picaxe programming cable** #PGM-08313 from SparkFun
- **LED holders (2)** #276-079 from RadioShack

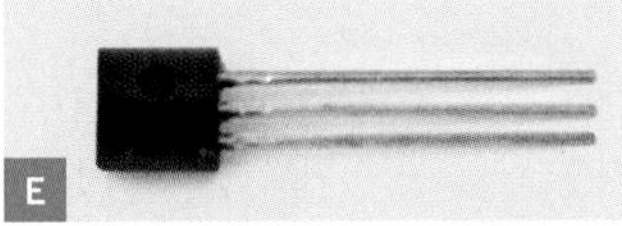

- **Knobs, to fit a 1/4" (6.35mm) shaft (2)** I used part #EL-HK7009 from Little Bird Electronics (littlebirdelectronics.com), or try #274RN-120L from RadioShack.
- **Wood boards, 3/8" or 8mm thick** You'll cut 7 small pieces; the exact dimensions in standard and metric measurements are given in Step 2. I used pine; you can use 3/8" pine craft boards, such as #50226 and #50234 from Lowe's (lowes.com).
- **Small box hinges, open width of 5/8" or 12mm (2) with matching self-tapping screws (8)** available packaged together from most hardware stores

TOOLS

- **PC computer, Windows or Linux**
- **Drill bits: 5/16", 1/8", 3/16", 1/4", 23/64", 15/32", 1/2"**
- **Wood glue**
- **Soldering iron**
- **Solder**
- **Drill**
- **Pencil**
- **Handsaw** for wood. Power saws can be used, but exercise caution because the pieces are small.
- **Wire cutters**
- **Screwdriver** to fit hinge screws
- **Pliers**
- **Chisel**
- **Beeswax or wood polish**
- **Sandpaper and sanding block**
- **Carpenter's square**
- **Wood clamps (2)** F-style, minimum 6" capacity
- **Ruler, standard 12" with metric**

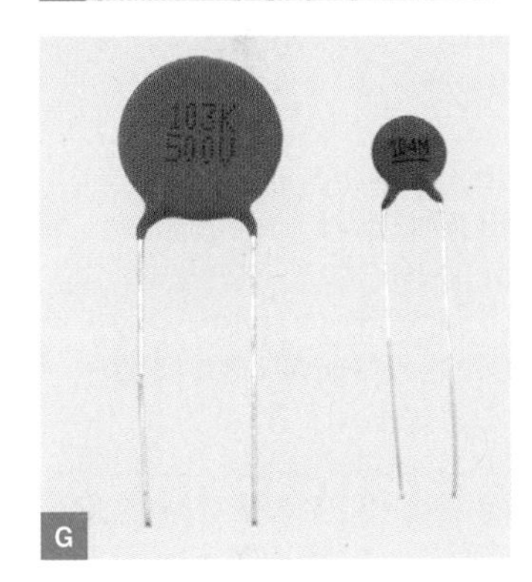

- **Allen wrench or screwdriver** used to fit grub screws (setscrews) on knobs
- **Hot glue gun and glue**
- **Computer with internet connection**
- **Hacksaw** if needed to cut stripboard to size
- **Sash clamp or band clamp** for gluing together side and end pieces of box
- **Headphones or amplifier with 1/4" audio jack** for listening to your new Luna Mod

REQUIRED RESOURCES

Download the program and schematic at makezine.com/26/lunamod.

- **Luna Mod BASIC code** *Luna_Mod.bas*
- **Schematic diagram** *Luna_Mod_Schematic.bmp*

Photography by Sam Murphy

MAKE IT.

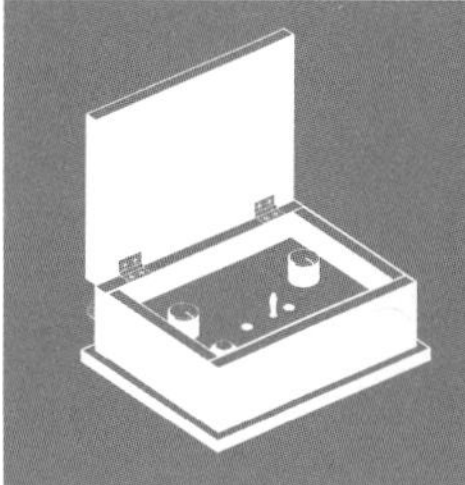

BUILD YOUR LUNA MOD LOOPER

Time: 1–2 Weekends
Complexity: Moderate

1. POPULATE THE CIRCUIT BOARD

1a. Take a 26×16-hole piece of stripboard (use a sharp hacksaw as necessary to cut to size, then sand the edges smooth). The copper strips should run the long way.

1a

1b. Cut, strip, and solder 13 solid-core jumper wires between the following points: (5,1 to 5,3), (5,15 to 5,16), (7,10 to 7,11), (7,13 to 7,15), (11,5 to 11,11), (13,5 to 13,7), (14,4 to 14,9), (16,8 to 16,14), (17,9 to 17,11), (19,10 to 19,12), (19,14 to 19,15), (21,9 to 21,14), and (22,8 to 22,13).

On the shorter jumpers you can remove all the insulation.

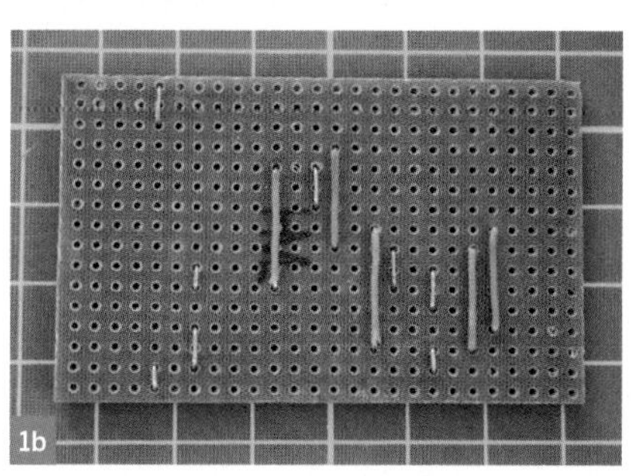
1b

1c. Turn a sharp ⅛" drill bit by hand to make breaks in the board's copper traces at these points: (5,12), (8,3), (8,11), (8,13), (11,3), (14,11), (14,12), (14,13), (14,14), (16,3), (17,14), (17,16), (19,8), (20,12), (24,6), (24,10).

Remember to count the *x* coordinate from right to left when the board's copper side is facing you.

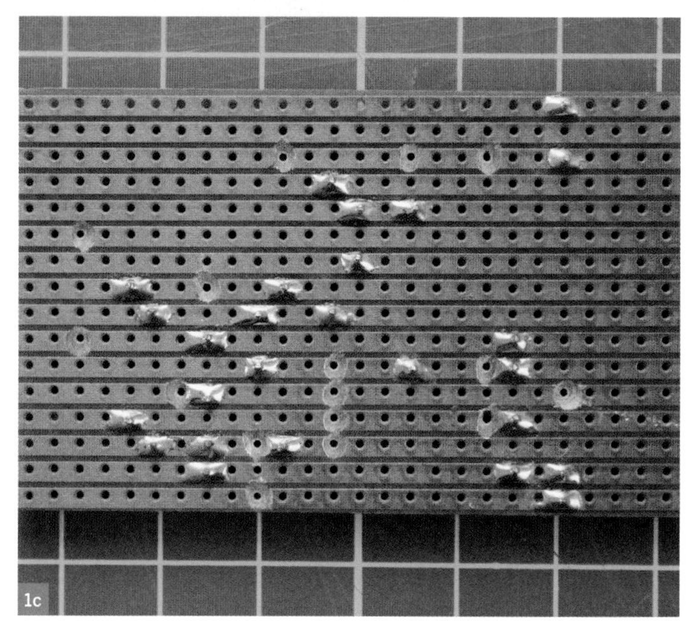
1c

NOTE: To identify holes on the board for placing components, I use grid references of the form (x,y), where (1,1) is the top left hole of the non-solder pad side, and the board's long edge and copper strips run along the x-axis.

For a schematic diagram of the Luna Mod circuit, see makezine.com/26/lunamod.

Photography by Brian McNamara

1d. Solder the resistors onto the stripboard: 1kΩ (18,12 to 18,16) and (25,12 to 25,16); 10kΩ (2,12 to 2,16), (9,7 to 9,14), (10,3 to 10,13), and (21,3 to 21,7); and 22kΩ (3,12 to 7,12).

1e. Solder in the 8-pin IC socket with pin 1 at (12,11) and pin 8 at (15,11).

1f. Solder in the 2 caps: 10nF (23,12 to 23,14) and 100nF (20,14 to 20,16).

1g. Solder in the voltage regulator, pin 1 to (3,5) and pin 3 to (3,3).

1h. Trim the LED leads to ½" (14mm) to keep them long enough to poke up through the panel, but keep track of which legs are positive (anode, with the longer leg) and which are negative (cathode, marked by a flat part in the plastic).

Solder one LED with its (+) leg at (9,3) and (−) leg at (9,4), and the other with (+) at (18,3) and (−) at (18,4).

1i. The on/off toggle switch legs are too wide to fit directly into the PCB holes, so use a ⅟16" drill bit to extend the holes into small slots at (14,1), (14,2), and (14,3).

Fit and solder the switch into place.

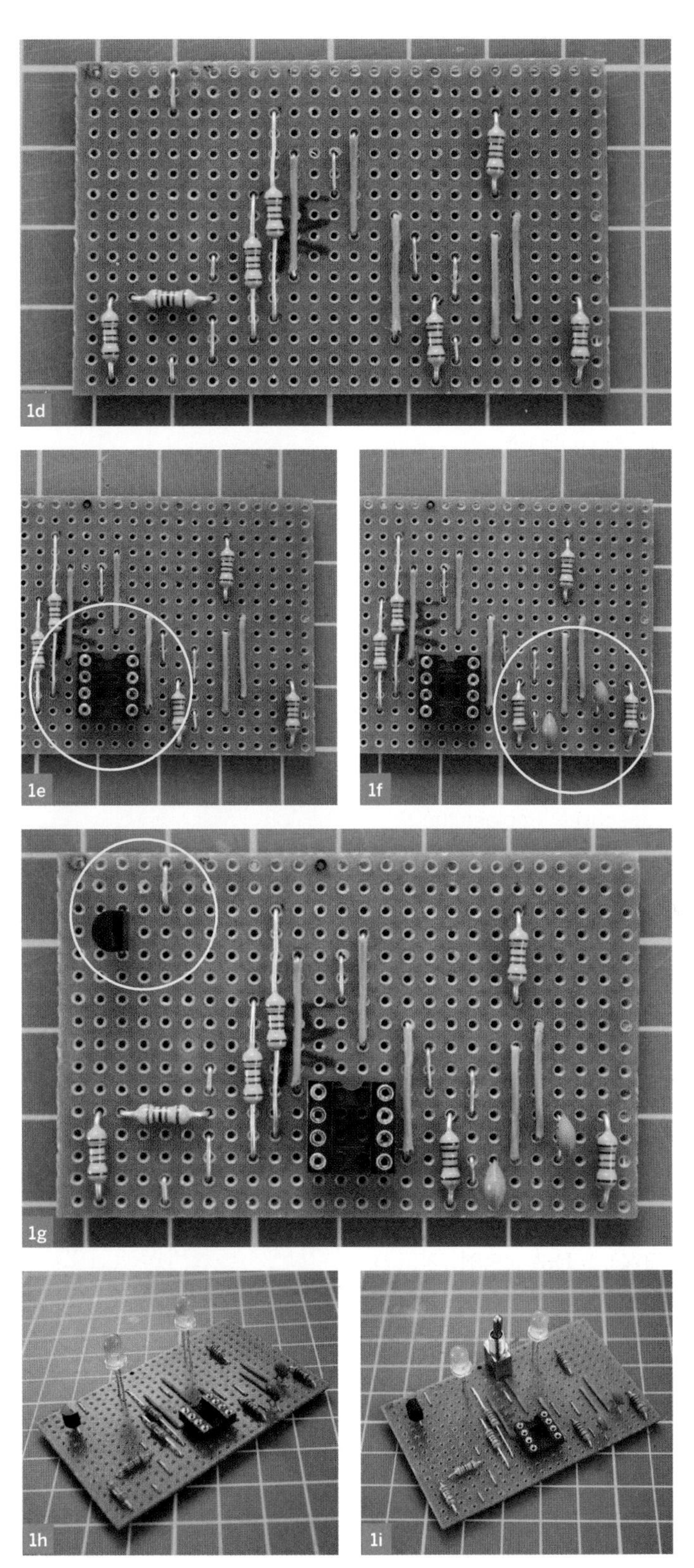
1d, 1e, 1f, 1g, 1h, 1i

1j. For the potentiometers, use the 5⁄16" bit to enlarge the holes at (1,6), (1,10), (26,6), and (26,10). Locate pin 1 on each of the pots; it's the right-most pin as you face the pins with the pot shaft pointing up.

Fit the pots' mounting tabs into the enlarged holes with pin 1 of one at (4,9) and pin 1 of the other at (23,7). Solder into place.

1k. For the programming socket, cut 3 wires, ideally of different colors, about 3½" (or 9cm) long. Solder one end of each to (1,11), (1,12), and (1,13). Remove the ⅛" stereo jack cover, slip it over the 3 wires, and solder the wires to the jack's contacts for the plug's sleeve (inner), ring (middle), and tip (end), respectively. Replace the cover.

1l. Solder 2 wire leads for the pushbutton about 7" (or 18cm) long at locations (1,14) and (1,15).

1m. Solder 2 wires about 7" (or 18cm) long to the PCB at (26,12) and (26,14). Solder the other ends to the panel-mount mono audio jack's tip (signal) and sleeve (ground) contacts, respectively.

1n. Solder the 9-volt battery clip to the board, red (+) to (1,2) and black (–) to (1,4).

1o. Plug the Picaxe 08M into the IC socket with pin 1 at upper left, nearer the LEDs.

1j

1k

NOTE: You'll solder the pushbutton to these 2 loose wires after fitting it into the front of the panel.

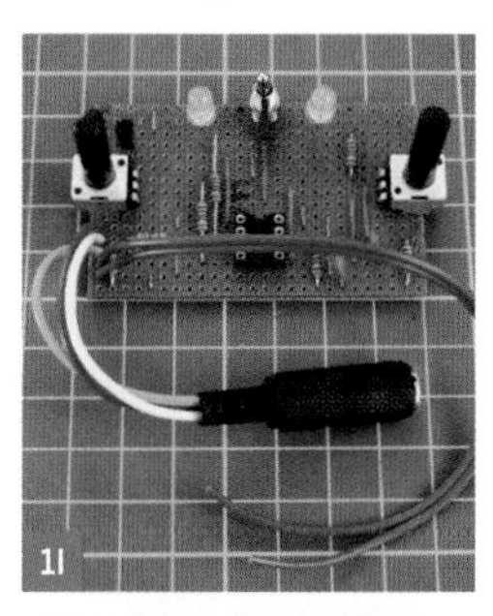

1l

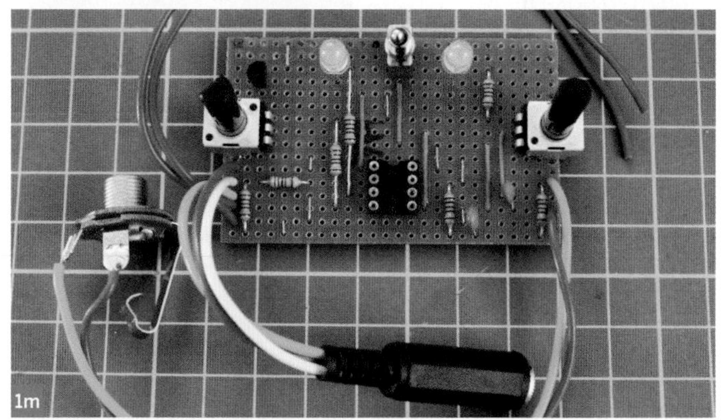

1m

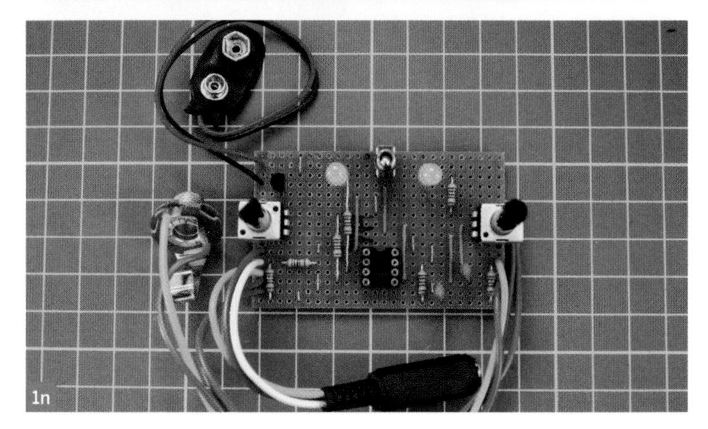

1n

The circuit board is now finished and ready to fit into the control panel.

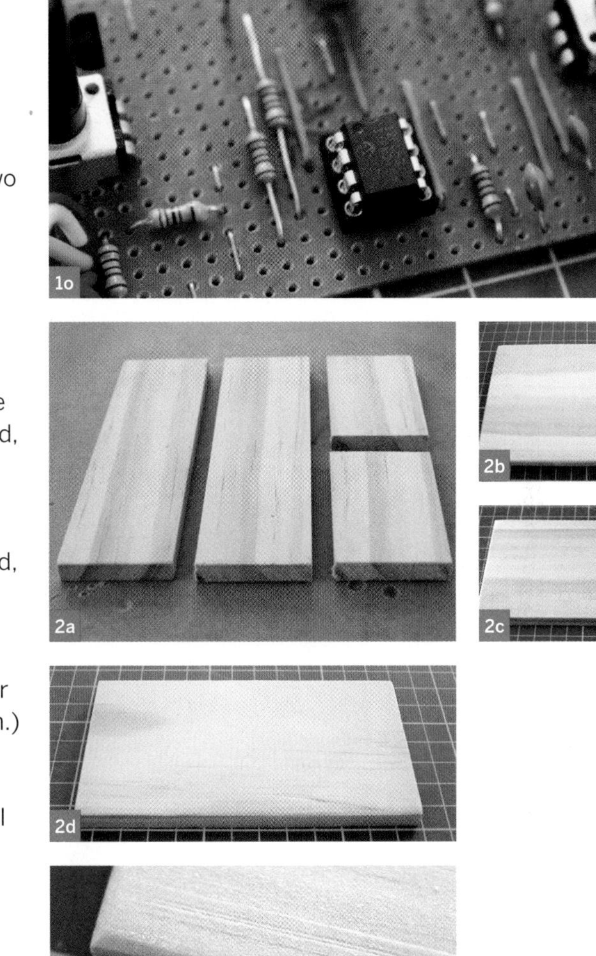

1o 2a 2b 2c 2d

2. BUILD THE CASE

2a. Using ⅜" boards, cut two 5½"×1⅝" side pieces and two 2½"×1⅝" end pieces. (Using 8mm boards, cut pieces 140mm×40mm and 65mm×40mm.)

2b. For the base, cut a piece 5⅞"×3⅝" (or for 8mm wood, 150mm×90mm.)

2c. For the lid, cut a piece 5½"×3¼" (or for 8mm wood, 140mm×82mm.)

2d. For the control panel, cut a piece 4¾"×2½" (or for 8mm wood, 124mm×65mm.)

2e. Use sandpaper and a sanding block to put a small bevel around the top edges of the base and lid pieces.

2e

2f. Assemble the sides and ends. Using a sash clamp or band clamp, glue and clamp the pieces together into a box shape, with the end pieces abutting the sides. Let dry.

2f

2g. Drill a hole in the bottom right rear corner of the enclosure for the audio jack. I used a 23/64" bit 20mm from the side and 8mm from the bottom, although anything in this general range will do. Countersink the hole slightly with a bigger drill bit.

2g

2h. Using wood clamps, glue and clamp the sides of the case centered onto the base piece. Let dry.

2i. Attach the lid. Place the hinges on the lid about 3/8" (or 1cm) in from the side, mark the screw holes with a pencil, and drill 1/16" holes for the hinge screws.

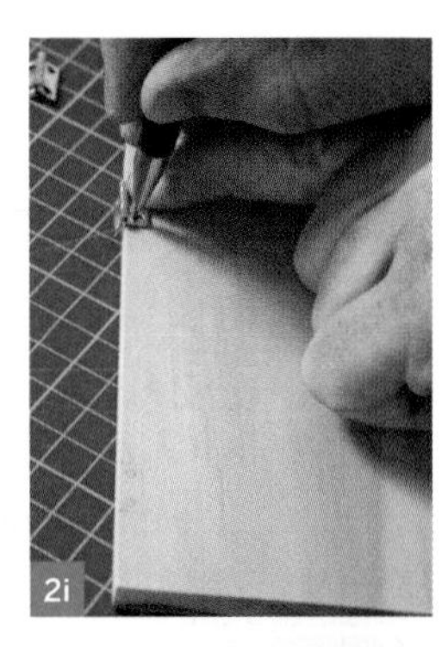
2i

2j. Screw the hinges to the lid.

2k. With the lid in place, use a pencil to draw around the position of the hinges on the rear side piece.

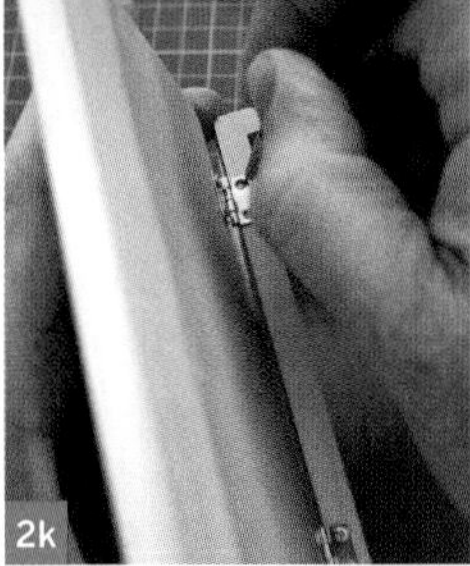
2k

2l. Use a small chisel to cut a recess into the wood in the shape of the hinges about 1/8" (3mm) deep.

2l

2m. Mark and drill 1/16" holes for the hinges in the recesses, and screw on the lid.

2m

2n. Drill the control panel. Using your completed circuit board as a template, drill the holes to fit in the control panel: 1/4" for pots and LEDs, and 3/16" for the toggle switch. Also drill a 15/32" hole at lower left for the pushbutton.

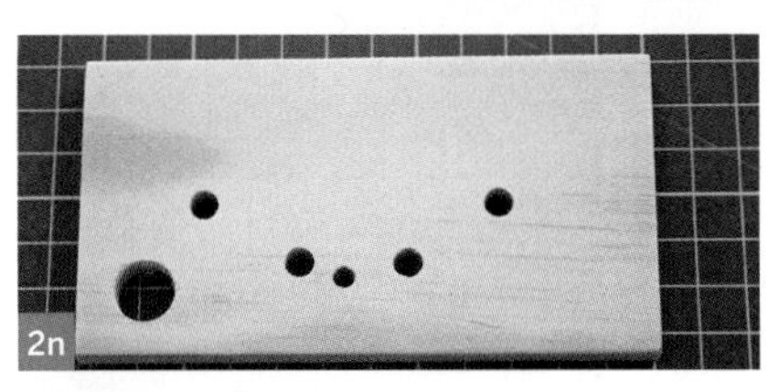
2n

2o. Finish the wood by rubbing it with beeswax or polish. Now the box is ready for fitting the electronics.

2o

3. ASSEMBLE THE HARDWARE

3a. Fit the plastic LED holders to the front panel. To help the LEDs fit, I snipped the 2 longer holding clips off.

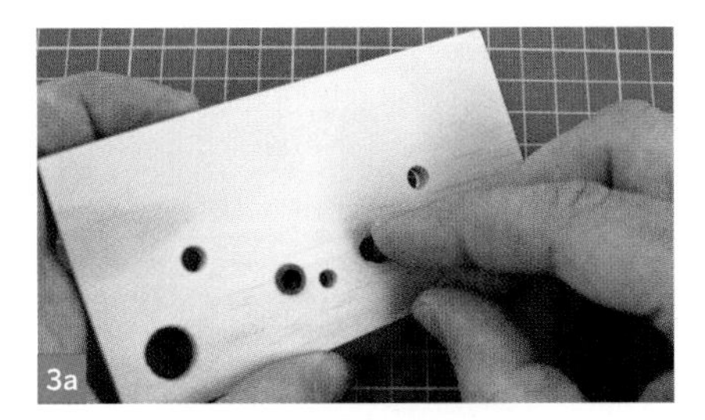
3a

3b. Press the circuit board onto the back of the control panel, fitting all components through their holes.

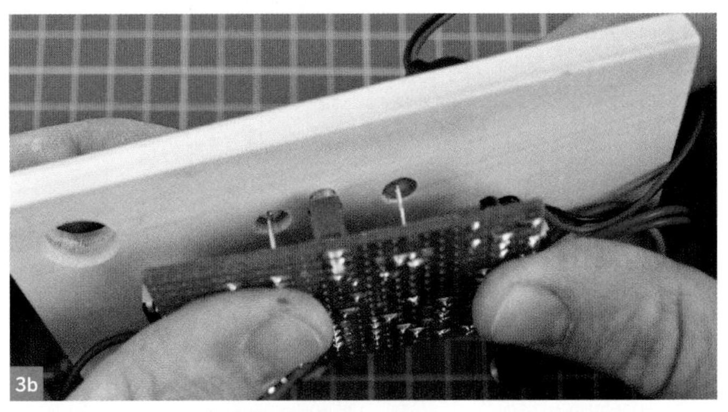
3b

3c. When the circuit board is in the right spot, tack the toggle switch to the wood with a little hot glue.

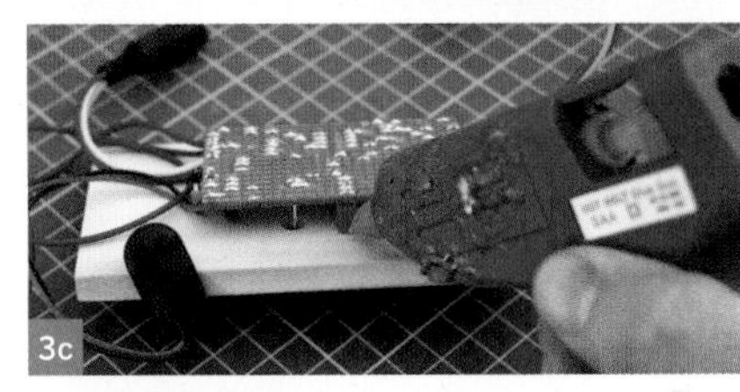
3c

3d. Fit the pushbutton switch through its hole in the panel front. Bend its solder tabs back, and solder the pushbutton switch wires to the solder tabs. Add hot glue to hold the pushbutton in place.

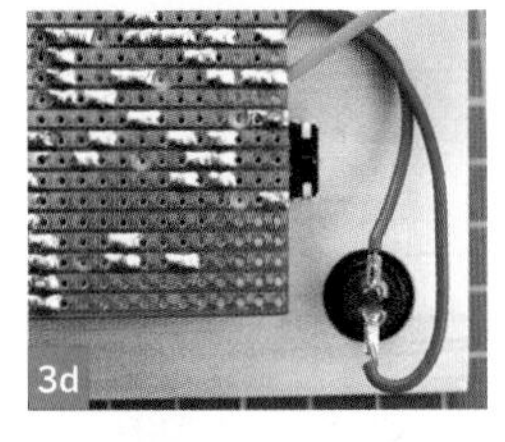
3d

3e. Fit the ¼" panel-mount audio jack to the case and hot-glue it in place.

3e

3f. Use hot glue to secure the audio jack programming port along the top edge of the circuit board.

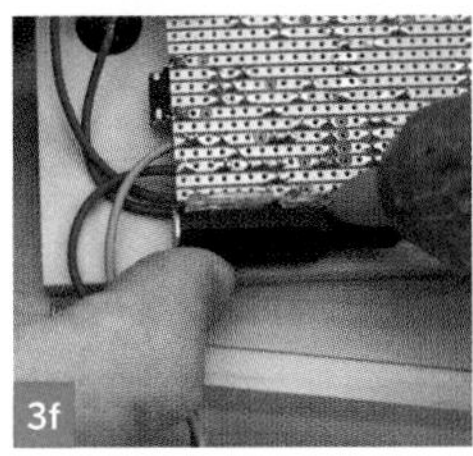
3f

3g. Clip a 9V battery onto the battery clip.

3g

3h. Hold the battery under the right side of the control panel, and fit the panel into the case.

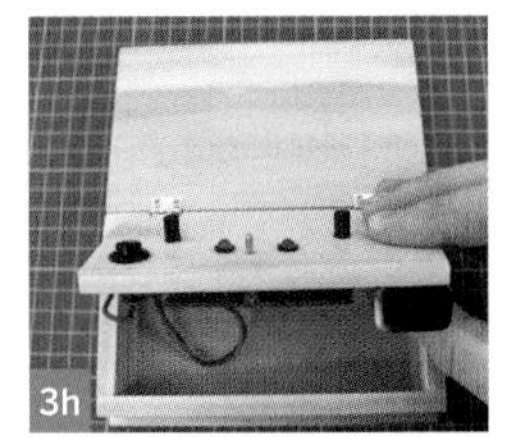
3h

3i. Screw the knobs onto the potentiometers with their included grub screws (set screws).

The hardware is now completed.

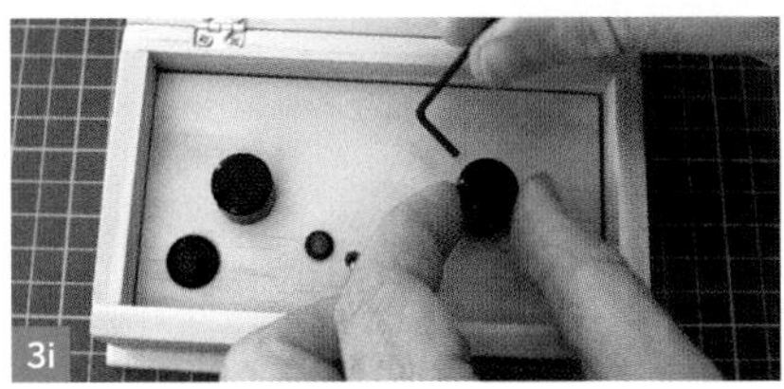
3i

4. PROGRAM THE LUNA MOD

4a. Download and install the free Picaxe Programming Editor Software from rev-ed.co.uk/picaxe, under the Software tab.

4b. Download the BASIC program file *Luna_Mod.bas* from makezine.com/26/lunamod.

4c. Lift up the control panel and connect the Picaxe programming cable between the Luna Mod programming port and your computer.

4d. Launch the Picaxe Programming Editor Software. Select File ⇒ Open, then select *Luna_Mod.bas* from the folder you downloaded it into.

4e. With the Luna Mod connected to the computer, power it up using its on/off switch.

4f. Press the computer's F5 key to compile and load the *Luna_Mod.bas* program onto the Luna Mod. After the program has transferred, you should see a "Download was successful!" message box. Then disconnect the programming cable and refit the control panel.

Your Luna Mod is ready to play!

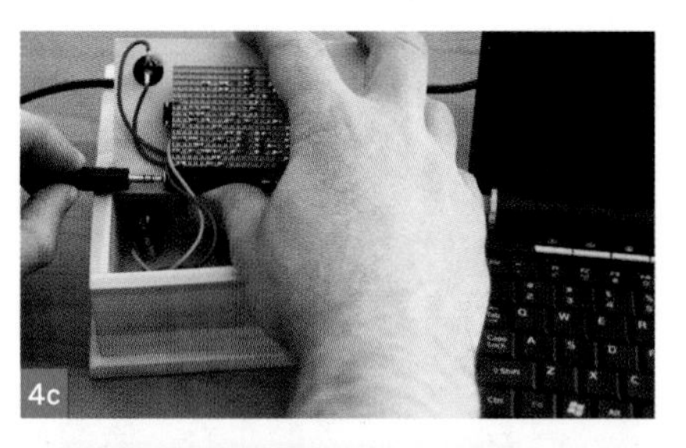
4c

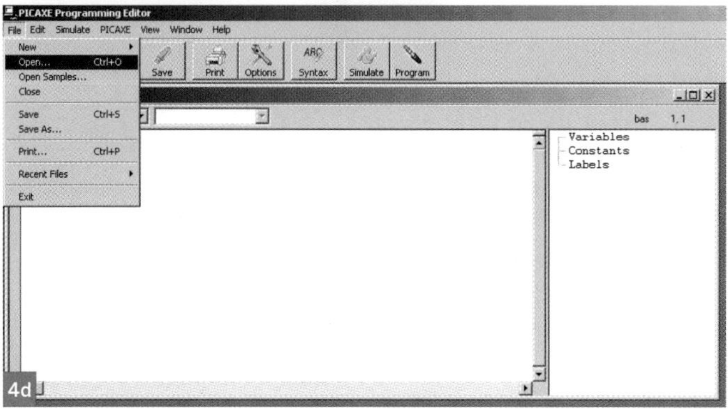

4d

4e

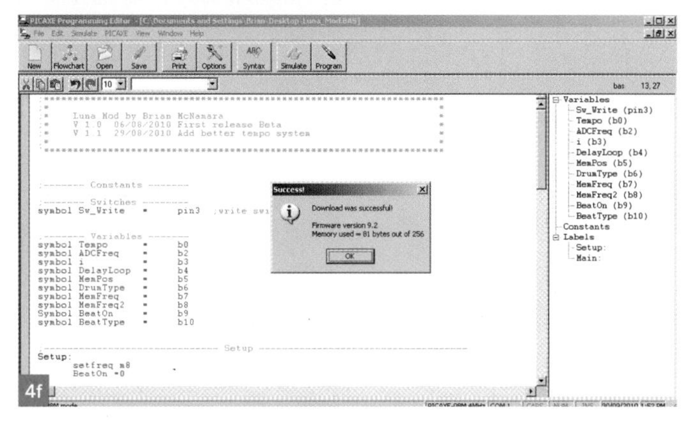

4f

USE IT.

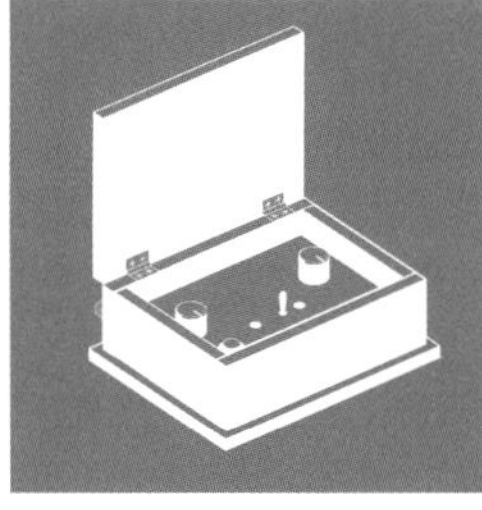

THE MOD SOUNDS OF THE LUNA MOD

Test, Play, and Have Fun

When you turn on the Luna Mod, the power light should come on, and the Tempo LED should flash. Plug some headphones into your Luna Mod and press the Write button — you will hear some noise. Turn the Sound knob while pressing the button, and the sound should change. Now move the Tempo knob — you should hear the loop getting faster and slower. If all that works, you're good to go. Otherwise, check your wiring against the Luna Mod schematic diagram at makezine.com/26/lunamod.

Playing the Luna Mod is a matter of experimentation. Keep in mind that you can always create breaks in the sound by turning the Sound knob completely counterclockwise.

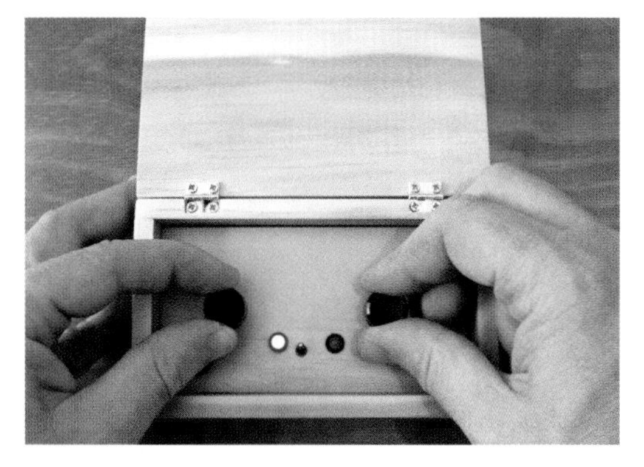

I had an enormous amount of fun learning how to play the Luna Mod because I'd never played an instrument like it before. I usually start with a basic rhythm that I set up by moving the Sound knob quickly clockwise while pressing the Write button. Then I add tones and blips by putting the Sound knob into various spots and quickly pressing the Write button once or twice. With this basic rhythm in place, you can let it repeat or keep adding and overwriting tones as desired.

Another trick for playing the Luna Mod is to record a loop with the Tempo knob at its slowest speed, then increase the tempo for a fast, highly detailed loop.

When I set up the Luna Mod for recording, I usually connect it directly to my mixer/recorder. For a more refined sound, I sometimes plug the Luna Mod into a reverb pedal, then a delay pedal, then the mixer/recorder.

Mod the Luna Mod!

The Luna Mod program only uses 81 of the 256 bytes available on the Picaxe 08M's onboard memory. So with the Luna Mod's built-in programming port, you have lots of room to modify and hack the software to make new sounds and functions. Enjoy, and let me know what you come up with!

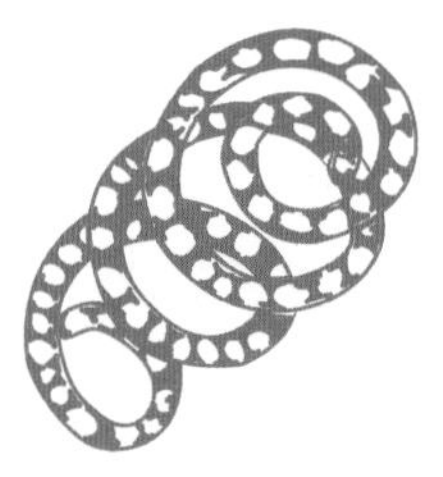

SPIRULINA SUPERFOOD TANK

Farm-Fresh Cyanobacteria

By Aaron Wolf Baum

Supplement your daily diet with fresh spirulina "superfood" grown indoors next to a sunny window.

In 2006, I worked with the Chlorophyll Collective at Burning Man to demonstrate how algae can capture greenhouse gases from generator exhaust.

Inspired by the power of these simple, single-celled plants, I then built a community algae lab in a shipping container in Berkeley, Calif. Ever since, we've been refining techniques for the home farming of spirulina (*Arthrospira platensis*), an edible algae that's well known for its health benefits.

This project presents our favorite DIY design: a home algae tank you can build in just a couple of hours after a trip to the aquarium store.

Algae farming is a revolutionary form of agriculture that can produce copious food, fuel, and other products in small spaces. It requires no land, and can even clean up air and water.

If you've eaten dried spirulina powder, which is sold commercially as a dietary supplement, you know it has an unappealing "seaweed-y" taste. But live spirulina has a lovely, creamy texture and a fresh, very mild flavor — really, almost no taste at all. Try it!

Dr. Aaron Wolf Baum received his training at Harvard and Stanford universities. He now teaches workshops on algae cultivation and sells kits for growing spirulina at home.

Photograph by Sam Murphy

SET UP: p.95 | **MAKE IT: p.96** | **USE IT: p.101**

Primordial Soup

You can grow algae in a regular aquarium; the trick is to fill it with a growth solution that nourishes spirulina while preventing colonization by other species.

(A) Sunny window admits light.

(B) Transparent container holds water-based growth solution (aka nutrient solution).

(C) Aquarium heater maintains ideal temperature for algae growth.

(D) Air pump and **(E) diffuser** bubble the water solution to prevent the algae from clumping, ensure that they receive similar amounts of light, and help supply carbon dioxide and remove oxygen from the solution.

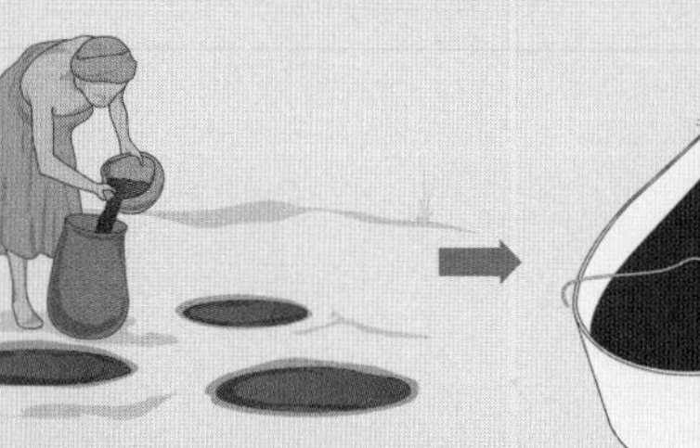

Timeless Simplicity

Taxonomically, spirulina are single-celled, prokaryotic (lacking a nucleus) algae of the type cyanobacteria, the organisms that put the first oxygen into the Earth's atmosphere and evolved into the structures that perform photosynthesis inside more complex, multicellular plants and algae.

Some natural lakes have long traditions of spirulina harvesting by indigenous people. On Lake Chad, in Africa, spirulina blooms are still harvested in clay pots, poured onto piles of sand to drain the water, then sun-dried into cakes.

The United Nations-affiliated Intergovernmental Institution for the use of Micro-algae Spirulina Against Malnutrition (IIMSAM, iimsam.org) promotes spirulina cultivation to combat hunger worldwide.

NASA has researched spirulina as a possible growable food for astronauts, and at its Ames Research Center, the agency is currently developing rooftop algal "raceways" for the mass culture of spirulina, and nanosatellites for investigating the effects of microgravity on algal photophysiology.

WHY SPIRULINA?

You can grow many types of algae at home. So why spirulina?

Safety. Spirulina can grow in extremely alkaline water that almost nothing else can survive. This makes it relatively safe.

Filterability. Spirulina's corkscrew shape lets you trap it with a simple filter, unlike spherical algae types that require centrifugation.

Nutrition. Spirulina supplies protein, vitamins, minerals, and fatty acids. Eating just half a gram per day can benefit your health.

Illustration by Alison Kendall

SET UP.

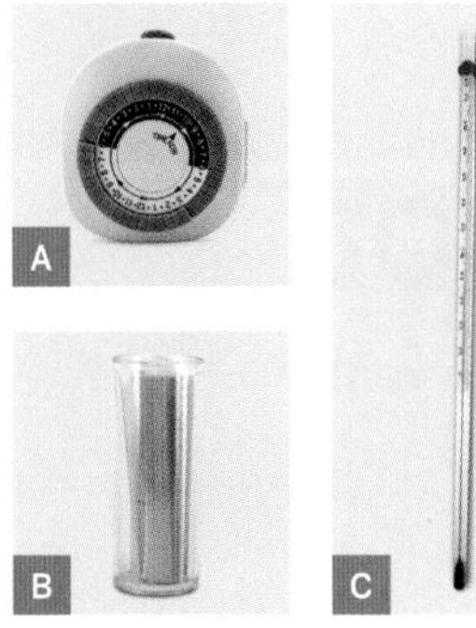

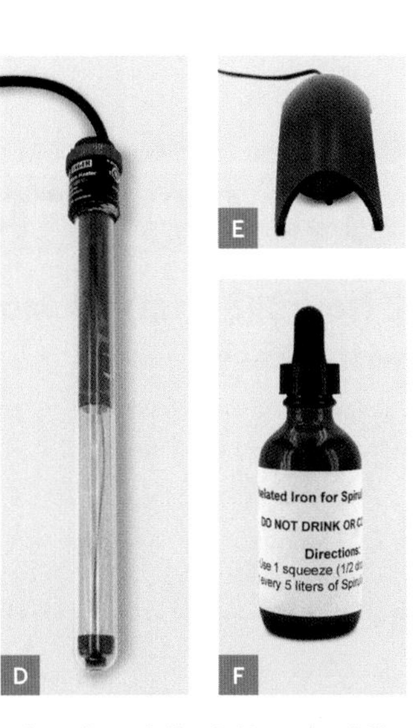

MATERIALS

Except where indicated, items are available from most aquarium stores.

A. Plug-in appliance timer available at most hardware stores for about $6

B. pH test paper that can test up to at least 11pH (yes, it must "go to 11") from a science supply company, such as part #3110J15 from Thomas Scientific (thomassci.com), $5

C. Aquarium thermometer one that's held by suction cups (without touching the glass) and reads up to 100°F

D. Aquarium heater powerful enough to take your tank up to a minimum of 94°F, such as the Eheim Jager 100W

E. Air pump on the upper end or over spec for your tank size. My usual bet is the Tetra Whisper Air Pump 10, but just about any aquarium air pump will do.

F. Chelated "Iron Juice," 2½ eyedroppers full AlgaeLab, $15 for 2oz, or mix your own using the supplied recipe and Steps 2b and 2c

NOT SHOWN

» **Transparent container** ideally a glass tank almost as tall and wide as your window but narrow in the direction perpendicular to the window so enough light reaches through. We used a 10-gallon tank, so the specifications and quantities given match this size.
» **Air diffuser (aka bubble wand)** sized for your tank. I recommend the 48" Marineland Flexible Bubble Wand
» **Nutrient "Starter Mix" powder, 1 bag** available from AlgaeLab (algaelab.org) for $19, or mix your own using the supplied recipe
» **Aquarium air line tubing, about 3'**
» **Vinyl tubing, at least ¼" internal diameter, about 3'**
» **Water, to fill container** Algae will grow faster at first if you filter out chlorination, which can be done using most home water filters, such as Brita. You can also use distilled water.
» **Live spirulina starter, about 1 liter** AlgaeLab, $49, or obtain from a spirulina-growing friend. Also, see makezine.com/26/spirulina for how to buy a sample tube from a culture library and grow it into a liter (but this costs more).
» **Nutrient "Make-Up Mix" powder, 1 bag** AlgaeLab, $29, or mix your own using the supplied recipe
» **Sunny window**

TOOLS

» **Buckets, 2½ gallon or larger (2)**
» **Clothespins or small clips (4)**
» **Filter material** Screen-printing fabric with 40- to 50-micron openings (or 325 mesh equivalent) is best, but silk cloth or even a paper coffee filter will do.
» **Kitchen scale** that can measure down to 1 gram
» **Measuring cup**
» **Measuring spoons**
» **Clear plastic film**
» **Mixing bowl or container, small**
» **Erasable marker**
» **Sheer white fabric**

RECIPES

Starter Mix

For 10 liters of growth solution

» **Sodium bicarbonate (baking soda), 160g** available at the supermarket
» **Potassium nitrate (saltpeter), food grade, 20g** item #850-16 from the Ingredient Store (store.theingredientstore.com), $19/lb
» **Sodium chloride (salt), 10g** available at the supermarket
» **Ammonium phosphate (either monoammonium or diammonium) or monopotassium phosphate, 1g** item #7305A from Homebrewers Outpost (homebrewers.com), $2/2oz
» **Lime, calcium chloride, or plaster, 1g (optional)** if your water is very soft

Iron Juice

For 10 liters of growth solution

» **Iron sulfate** either supplement drops (such as Enfamil Fer-In-Sol), 1½ eyedroppers full; or in solid form, sold as plant fertilizer, 100mg
» **Green tea, brewed strong** 1 cup is more than enough.

Make-Up Mix

Used to replace nutrients at harvest time. See Starter Mix recipe for notes and sources. This makes enough for several months of harvesting.

» **Potassium nitrate (saltpeter), 1.4kg**
» **Ammonium phosphate or monopotassium phosphate, 50g**
» **Potassium sulfate, 30g** item #SB17028M from Nasco (enasco.com), $8 for 100g
» **Magnesium sulfate (Epsom salts), 20g** available at the supermarket
» **Lime, calcium chloride, or plaster, 10g (optional)** if your water is very soft

(Recipes adapted from "A Teaching Module for the Production of Spirulina," by J. Falquet, Antenna Technologies, June 1999.)

Photography by Sam Murphy

GROW SUPERFOOD

Time: 2 Hours (Setup) + 1 Month (Harvest)
Complexity: Easy

1. SET UP YOUR TANK

1a. Attach the aquarium air diffuser to the air pump using the air line tubing, cutting the tubing just long enough to let the diffuser lie in the bottom of the tank. If your diffuser has an anti-siphon valve, install that as well following the included instructions.

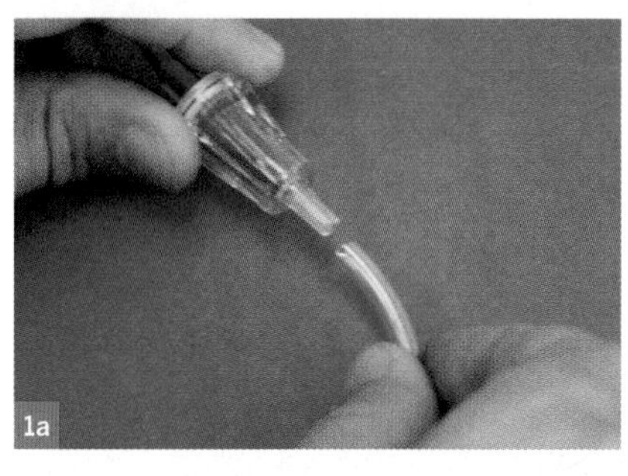

1a

⚠ CAUTION: Never plug in the heater unless it's fully immersed, and always unplug it before emptying the tank. Otherwise it may overheat and break.

1b. Bend the air diffuser to route it around the inside edges of the tank. This will help keep the walls clean and also let you see the rising bubbles, for a "mad scientist" effect.

Place the heater in the tank, but don't plug it in yet. You'll be starting with the tank only about one-quarter full, so position the heater down toward the bottom.

Place the tank in the sunniest window in your house and make sure it's easily accessible so you can add the nutrient solution.

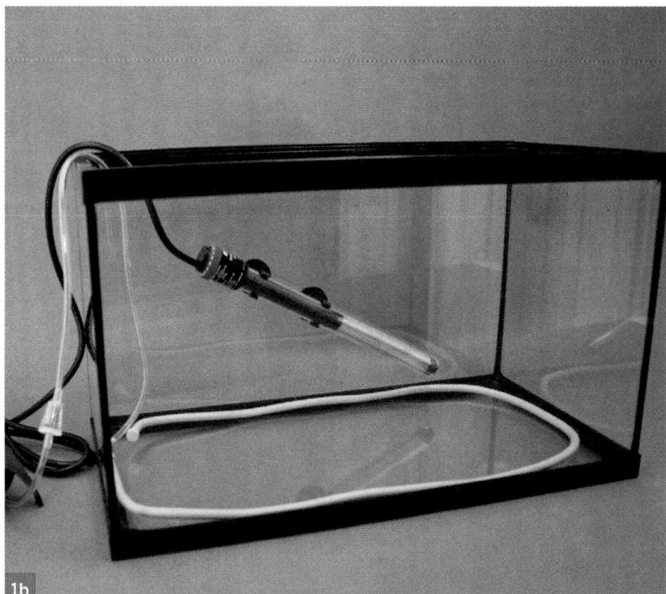

1b

2. MIX THE GROWTH SOLUTION

2a. In a large bucket, make 10 liters of nutrient solution by mixing 10 liters of fresh water with 1 cup of Starter Mix.

2a

Photography by Ed Troxell

You want enough to fill your tank or other container about a quarter of the way, so if you want to mix more or less, use 1½ tablespoons (4½ teaspoons) of powder per liter.

2b. To make your own Iron Juice, brew a strong cup of green tea. When the tea is a deep green color (it doesn't need to be hot), measure 2½ teaspoons (or ¼ teaspoon per liter of solution) into a small bowl or container.

2b

2c. If you're using liquid iron sulfate, measure and add 1½ eyedroppers full into the tea. With solid iron sulfate, grind 100mg into a fine powder and add it to the tea. Swirl it around to mix thoroughly.

2c

NOTE: Notice that the tea turns purple at Step 2c. This is due to a process called *chelation*, by which tannins in the tea form iron-cored rings, a molecular structure that tends to absorb visible light. Similar structures give hemoglobin (the red in blood) and chlorophyll (the green in plants) their color. This process is very similar to how ink was made up until modern times — check out "iron gall ink" on Wikipedia.

2d. Now add the iron sulfate-green tea mixture (aka Iron Juice) to the nutrient solution in the bucket and mix well. A small amount of Starter Mix powder may remain undissolved, which is fine. Pour the solution into your tank.

2d

3. PLANT THE ALGAE

3a. Add the live spirulina to the tank.

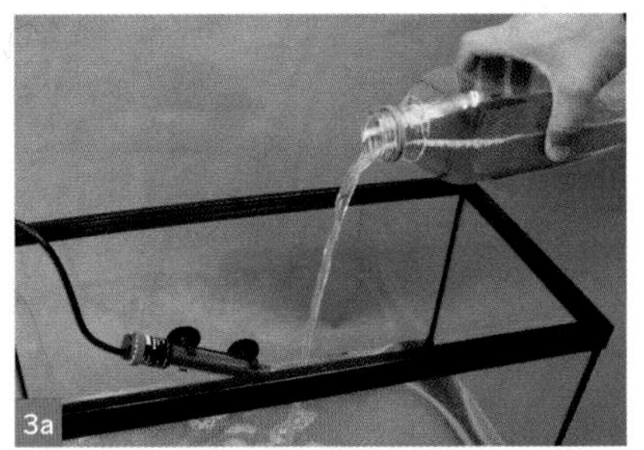
3a

3b. Plug the heater in and turn it up to 98°F, or a minimum of 90°F. Growth will be notably reduced below about 90°F, and above 100°F the heat will stress the spirulina.

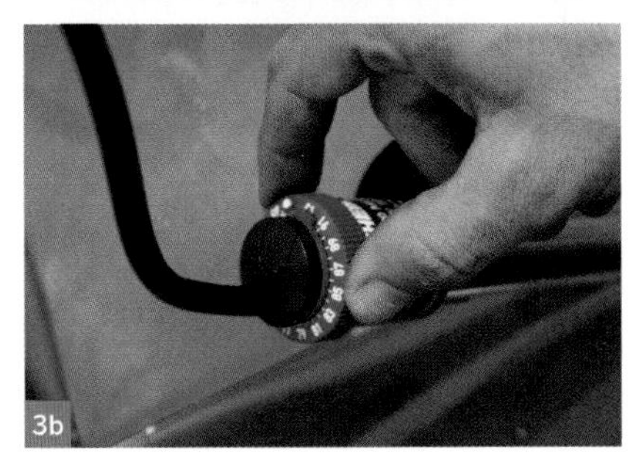
3b

NOTE: With the Eheim Jager heater, read the instructions for adjusting it to achieve these temperatures, which are warmer than usual aquarium settings.

3c. Angle the heater so that as much of its submergible element as possible is in the solution. Again, glass is recommended, but if your container is plastic, don't let the heater touch the wall — it might melt a hole!

3c

3d. Plug the timer in nearby and set it to turn on 3–4 hours before sunrise and turn off at sunset. Plug the heater into the timer, and plug in the pump.

3d

3e. Cover the tank with clear plastic to reduce evaporation, retain heat, and keep out flies. Then mark the tank depth with an erasable marker.

3e

NOTE: For the first week or so, the algae may be overwhelmed by all the sunlight (if they are stressed by this or anything else, they acquire a yellowish color). If this happens, move the tank away from the window a bit or shield the algae from direct sun with sheer white fabric. After they have thickened up a little, give them as much light as possible.

4. WATCH IT GROW!

4a. Monitor the temperature in the tank. It should be in the mid- to high-90s during the day, and at least in the 80s at dawn. Don't let it exceed 102°F. If it gets too hot, shade the culture with sheer white fabric and temporarily remove the plastic cover to allow some cooling evaporation.

Every few days, add fresh filtered (or distilled) water to the tank to keep the level of liquid up to the mark.

4b. After 2 weeks, the algae will have thickened up. It's time to double the volume (or top it off). Mix in another batch of growth solution.

Mark the new level on the tank and maintain it there by adding water as needed.

4

Photography by Aaron Wolf Baum (Figures 4 and 5)

4c. Repeat the process — waiting until the algae have thickened, then expanding the culture — until you have a full tank of thick green stuff. When it reaches 3cm density (see "Harvesting and Maintenance"), be proud of yourself. It's time to harvest and enjoy!

5. HARVEST THE ALGAE

5a. Use small clips or clothespins to stretch the filter material across the mouth of the bucket. Let it droop down in the middle, but make sure the edges are as high as possible.

5b. Decide how much you want to harvest; a quarter to half of the tank is typical. Put the bucket next to the tank so that the lowest point of the filter sits the chosen amount below the top of the liquid.

5c. Time to siphon! Place one end of the ¼"-diameter vinyl tubing into the culture, and put the other end in your mouth. Breathe out (you'll blow some bubbles), then suck the culture up the tube, just shy of your mouth. Pull the tube out of your mouth and immediately cover the end with your thumb.

You should now have a tube full of green stuff (this may take a few tries). If you get some culture in your mouth, don't swallow it, but don't worry — it's not harmful. It will taste like baking soda.

5b

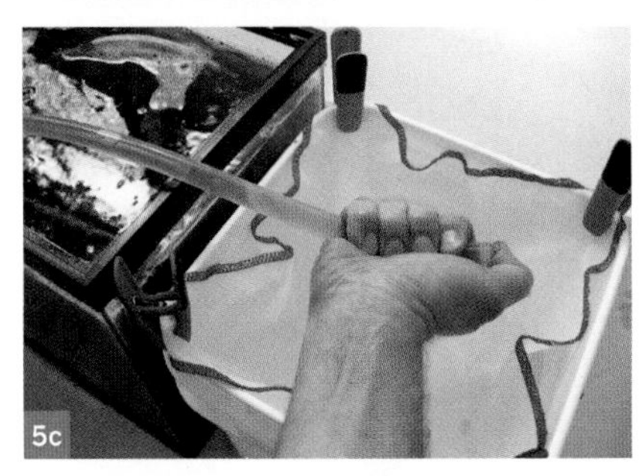
5c

TIP: You may need to rest the bucket on something lower than the tank base to harvest larger amounts.

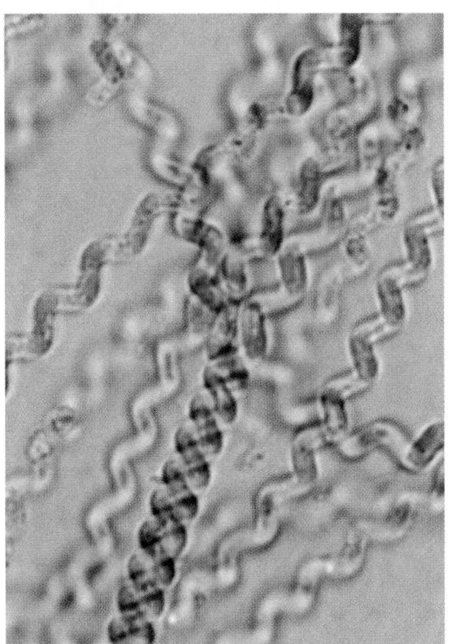

NOTE: I recommend looking at the spirulina under a microscope sometimes, not just to be sure that your culture is pure but also because it's really cool to see their little helices and reflect that they're your relatives.

Photograph Corbis (microscope)

5d. Center the tube end just over the filter and let go. The culture should flow through the tube into the filter. Try to avoid letting it overflow the filter's sides (you can raise the tube slightly if it's siphoning too fast, and try moving the tube end around the filter material). The idea is for all of it to pass through the filter. See the green stuff accumulating? That's your spirulina!

5e. When you've harvested the desired amount (or if the liquid level in the bucket is getting close to the filter, which is best to avoid), stop the siphon by raising the tube end up and letting the culture flow back into the tank.

5f. You now have a watery paste of spirulina on the filter. Raise the edges of the filter, gather the paste into its center, and gently squeeze it to release the paste's remaining liquid into the bucket. If a lot of green stuff comes through the filter, you're squeezing too hard. You should end up with a thick, dark blue-green paste. It's ready to eat!

5g. Estimate the number of tablespoons of drained algae you harvested. To replace its nutrients, add 1 teaspoon Make-Up Mix to the liquid in the bucket for every tablespoon of algae harvested. Also make and add a little of the Iron Juice as in Steps 2b–2d. Mix it up, then pour the liquid back into the tank.

You're done!

5d

5f

5g

⚠ CAUTION:
To ensure that the spirulina is safe to eat, use the pH test papers to test the pH of the liquid. Do not eat any spirulina from the tank if its pH is below 10. (There is no particular reason to believe that there's anything harmful in the culture, but it's the high pH that prevents other organisms from growing.)

Also, if anything else seems wrong, don't eat the algae, like if the color is off (it should be dark blue-green), if you see a lot of clumping, if the tank has an unpleasant odor, or if some object of questionable origin is in the tank, etc.

USE IT.

ROLL IN THE GREEN!

Harvesting and Maintenance

Wait until the tank reaches at least 3cm density before harvesting — in other words, when you can no longer see a white object a few centimeters in diameter through 3cm of the liquid.

If the tank is well maintained and in a lot of direct sunlight, you should be able to harvest spirulina every few days. Once you're harvesting regularly, add a shot of "Iron Juice" — a pinch of iron sulfate in a shot glass of green tea — to the tank every couple of weeks.

Using these methods should keep your tank going for several months. Over time, however, the algae will push up the pH, and eventually you'll notice their growth slow down. The algae may become a bit yellowish (when they're happy, they're a gorgeous blue-green) and may start clumping up and sticking to the walls.

These signs probably mean it's time to refresh the growth medium (although they can also happen for other reasons, such as high temperature or a nutrient imbalance).

To refresh the medium, follow Steps 2a–2d again to mix 10 liters of fresh nutrient solution (or whatever amounts to a quarter of your tank) in a bucket. Move the heater to the bucket to start warming up the new medium. Follow the harvesting procedure, but drain your whole tank, putting the harvested spirulina into the new starter medium immediately; no need to squeeze.

Once the tank is drained, wipe it out, pour the starter culture back into the tank, and put in the heater, just as you did when you started out. Over the next week or so, expand the culture back up to the full tank, and you're back in business!

Photograph by Garry McLeod

Bon Appétit!

I often just scoop up the spirulina paste with a spoon and eat it directly; it has a nice buttery texture and almost no taste. You can also throw it in a smoothie (see craftzine.com/go/spirulina for recipes), mix it with guacamole, or put it on crackers with cheese. For more inspiration, check out the "Spirulina Cookbook" on algaelab.org.

Due to its high level of nutrients and digestibility, spirulina goes bad pretty fast, like raw eggs. If you don't plan on eating it right away, put it in the fridge or freezer. Fresh spirulina will last 3 days in the fridge and indefinitely in the freezer.

Remember, although people have been growing and eating spirulina without ill effects for a long time, we cannot absolutely guarantee the safety of growing and eating your own spirulina. Follow these directions closely, and don't eat it if you have any reason to suspect that anything is wrong!

Most importantly, join the international spirulina home-grow network. Post your results and questions on the algaelab.org forums, and I'll do my best to answer and give you props for your ideas and accomplishments.

PRIMER

BIOSENSING

Track your body's signals and brain waves and use them to control things.

By Sean M. Montgomery and Ira M. Laefsky

As we engage with the world, our bodies react with unconscious signals that until recently have been measured almost exclusively in doctor's offices and research labs: nerves fire, eyes saccade, hormones release, etc. But the past decade has seen an explosion of inexpensive biosensors and bio-enabled products ranging from cardio watches to brain wave-controlled video games.

Tools to measure such phenomena as heart rate, brain waves, blood pressure, skin resistance, and even facial expressions are now inexpensively available to DIYers.

The possibilities of jacking into these bio-signals seem virtually endless, verging on science fiction. By monitoring some phenomena (biofeedback) you can train yourself to modulate them, possibly improving your emotional state. Biosensing lets you interact more naturally with digital systems, creating cyborg-like extensions of your body that overcome disabilities or provide new abilities. You can also share your bio-signals, if you choose, to participate in new forms of communication.

The utility of biosensing applications should multiply as more people hook themselves in and network their signals. But this future raises questions: will Big Brother have access to your thoughts and feelings? Makers can guide biosensing's future not only by creating new applications but also by hacking commercial systems and providing oversight to enforce a balance between utility and privacy.

This article reviews the bio-signals that hobbyist-accessible equipment can read, then details two projects: the Truth Meter and the Brain Blinker. The Truth Meter, which is easily built on a solderless breadboard, detects increased levels of sweating (and therefore arousal) by measuring skin conductance.

The Brain Blinker shows how to use a $200 EEG headset to plot live brain wave readings and send them into an Arduino microcontroller. The Arduino displays the data on an LED bar, and can just as easily be programmed to perform other functions with it.

Sean Montgomery (produceconsumerobot.com) has a Ph.D. in neuroscience and works as an engineering consultant, new media artist, and entrepreneur in New York City.

In 1977, **Ira Laefsky** co-developed the first computer menu system based on eye-gaze tracking. He is a retired senior consultant for Digital Equipment Corporation and Arthur D. Little.

Photography by Cody Pickens

THE QUANTIFIED SELF

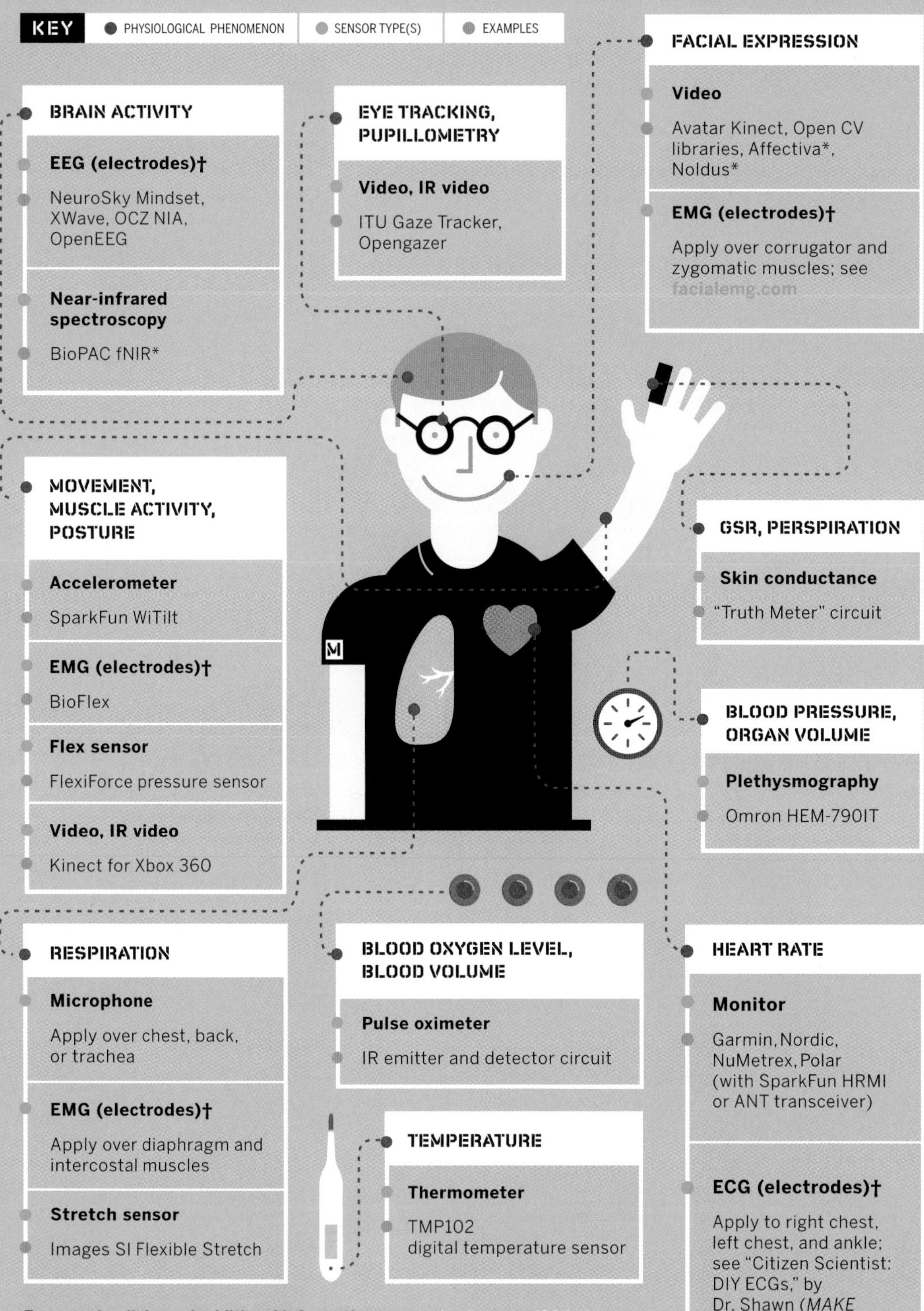

For sourcing links and additional information, see makezine.com/26/primer

* Items are beyond typical DIY budgets, at least for now.

† Peel-and-stick electrodes (available cheap in bulk) with alligator leads clipped to the metal nipples on the back.

TRUTH METER

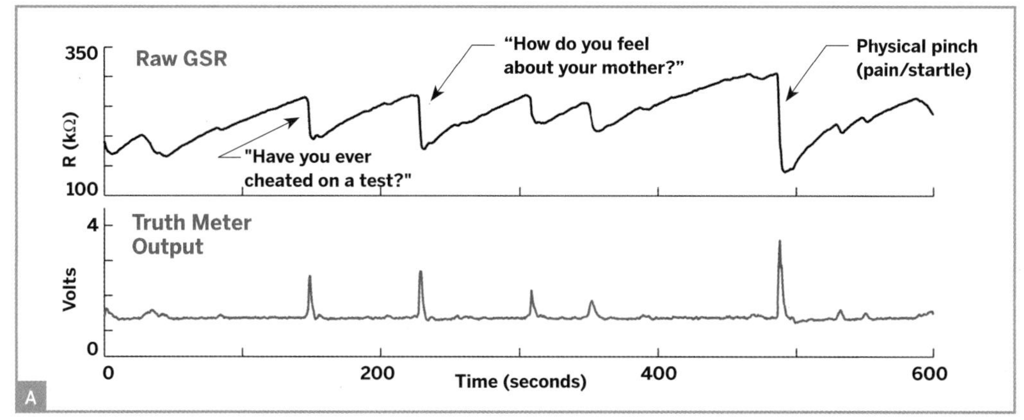

THE SWEAT RESPONSE

When you experience an arousing stimulus, like an evocative question, a startling noise, or even a disturbing thought, your body generates a variety of psychophysical responses.

One of these is micro-pulses of sweat released after a 1- to 2-second delay from apocrine sweat glands that are tied to the arousal systems in your body via adrenaline and other hormones. The reason your palms might get sweaty during public speaking or a job interview, for example, is because your mind is on high alert and every small stimulus generates one of these pulses.

Each pulse of sweat increases the electrical conductance of your skin, and when this conductance is measured and tied to arousing stimuli, it's referred to as galvanic skin response (GSR). The Truth Meter measures GSR for display on an LED or for input to a microcontroller. It's called a Truth Meter because GSR is an important component of lie detector (polygraph) tests used to assess how nervous subjects are while answering questions during interrogation.

So how do we measure GSR? The first step is to understand the signal. With each pulse of sweat, skin resistance decreases suddenly and creeps slowly back up as the sweat evaporates. The Truth Meter circuit transforms this pattern of drops and slow recoveries into sharp spikes deviating from a steady baseline, to light an output LED or trigger some other action. Figure A shows Sean's raw skin resistance and Truth Meter output recorded over a 10-minute period, during which friends asked him questions and pinched him.

The resistance sensor itself is simply 2 metal cuffs attached around your fingers with a piece of velcro (Figure B). Hypoallergenic metals used in jewelry and those less reactive with skin, such as stainless steel, are the best, but any solderable metal will work. We recommend copper or brass foil. Fingers are a good place for measuring GSR because apocrine sweat glands occur in very high concentrations on fingers and palms.

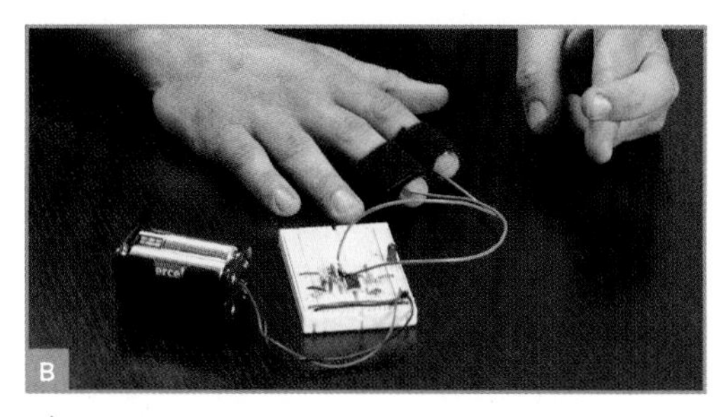

⚠ BIOSENSOR PRECAUTIONS

Human bodies are irreplaceable assets, so exercise care when working with biosensors. Here are some guidelines:

Avoid using AC (wall outlet) power, and electrically isolate electrodes and other parts in contact with the body from any high-voltage power source. Use batteries if possible, and an unplugged laptop instead of a desktop computer.

Sterilize electrodes and other biosensors with rubbing alcohol before use. Don't penetrate or abrade the skin when applying them or place them anywhere inside the body. Follow manufacturer's directions, and exercise reasonable common sense when handling and placing biosensors.

Diagrams by Sean Montgomery

MATERIALS

Velcro tape, ¾", 7" long
Copper or brass foil, 36 gauge (0.005") or thinner, 1" wide, 6" long available from hobby or craft stores
Insulated wire, 18 gauge, 10" lengths (2)
Headers, 2-pin (2) or male breakaway header cut into 2-pin segments
Solderless breadboard such as part #276-001 from RadioShack (radioshack.com), $20
Jumper wire kit RadioShack #276-173
Batteries, AA (4)
Battery holder, 4xAA
Dual op-amp IC, MCP6002 part #MCP6002-I/P-ND from Digi-Key (digikey.com)
Resistors, ¼W: 220Ω (1), 10kΩ (1), 100kΩ (1), 1MΩ (1) 3.3MΩ (2)
Capacitors: 10nF (1), 0.1μF (2)
Diodes, 1N4001 (3) RadioShack #276-1653
LED, red

1. MAKE THE 2 SENSORS

1a. Cut some ¾" velcro tape into a 3½" piece of loop (soft) side and a ¾" square of hook (rough) side (Figure C).

1b. Peel off the tape backing, and align and stick the hook velcro, back to back, at one end of the loop velcro.

1c. Cut a 10" length of 18-gauge wire and strip ½" off one end. Lay the stripped end of the wire perpendicularly across the end of the loop velcro's sticky side so that its stripped portion hangs off the edge (Figure D).

1d. Lay a 1"×3" strip of copper or brass foil over the sticky portion of the loop velcro and wire. Press together firmly and evenly and trim off the excess foil (Figure E).

1e. Fold the stripped portion of the wire back over the foil and solder it in place. Try to create a smooth, flat surface with the solder that won't dig into your finger.

Repeat 1a–1e for the second sensor.

1f. Strip ¼" of the sensor wire ends and solder them to a 2-pin header so they can be inserted into the breadboard (Figure F).

DISCLAIMER: The Truth Meter reveals changes in arousal, but it's not really a "lie detector." Even polygraph tests, which measure GSR, heart rate, blood pressure, etc., are not admissible as evidence in many courts.

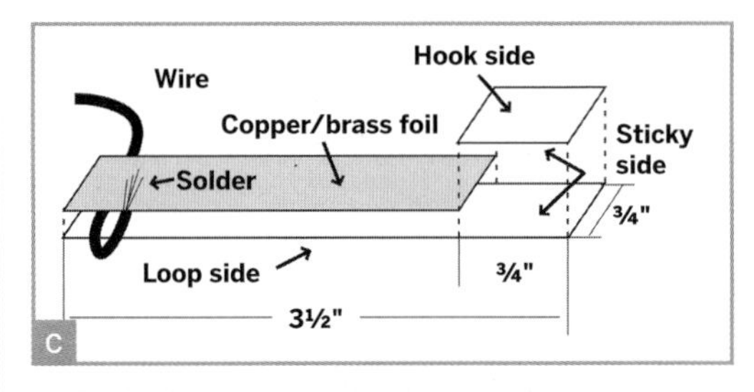

C

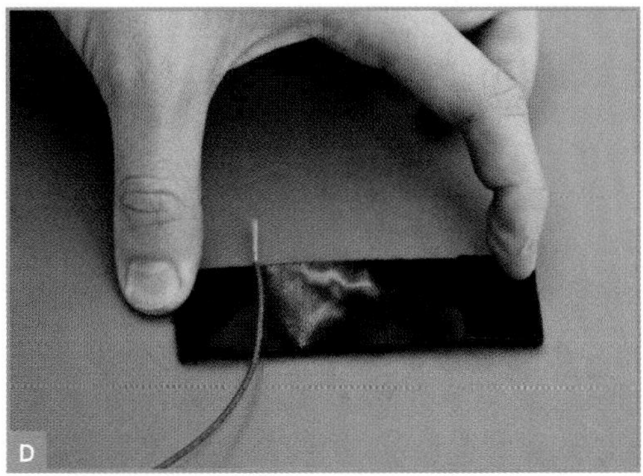
D

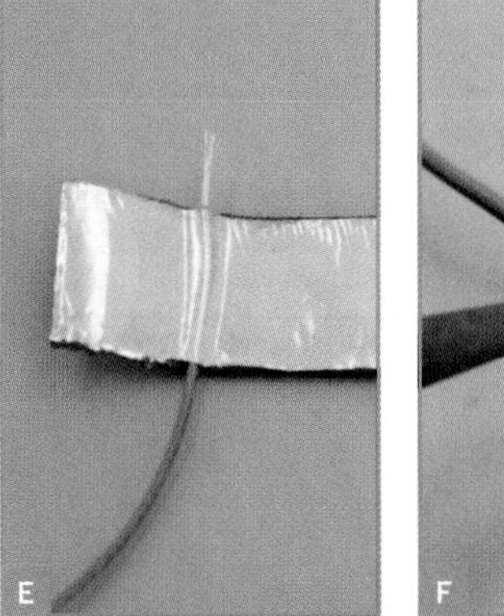
E

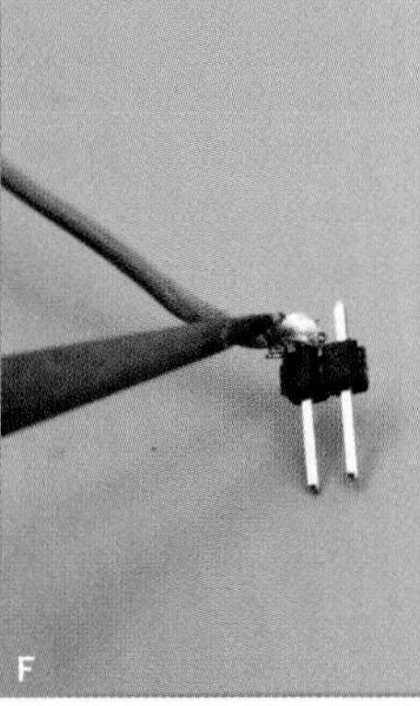
F

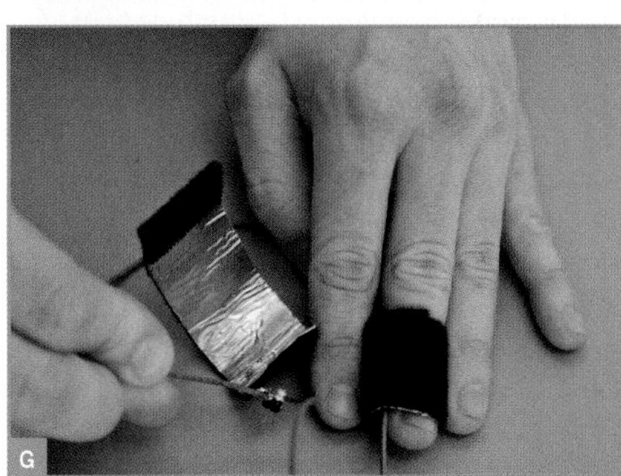
G

Photography by Sam Murphy

2. BUILD THE CIRCUIT

The Truth Meter circuit is quite simple. The schematic diagram in Figure H shows all its electronic components and connections, and Figure I shows them assembled using jumper wires on a standard 830-hole solderless breadboard from RadioShack (see makezine.com/26/primer for a higher-resolution photo). This section includes the "conceptual steps" for building it, following the GSR's signal path from sensor to amplified output.

2a. The battery voltage connects to the power rails along each side of the breadboard. The sensor cuffs connect to V+ and V–, with a high-resistance (1MΩ) resistor between V– and one cuff. This resistor acts as a voltage divider, converting the resistance from the sensor into a voltage. Using a high-value resistor this way can introduce noise, but it has the benefit of making the resistance-voltage curve roughly linear over a wide range of skin resistances.

2b. Skin resistance can vary greatly, depending on individual differences, weather, mental state, etc., so our circuit needs to calibrate to a steady baseline. To enable this, a resistor/capacitor (RC) high-pass filter cuts out longer frequencies under ~0.5Hz. This steadies the signal to create a usable baseline while still letting through the shorter GSR signals.

2c. Another RC circuit forms a low-pass filter to remove frequencies above ~5Hz, thus filtering out high-frequency noise such as 60Hz originating from nearby AC power wires.

2d. A series of diodes sets the baseline voltage (from V–) at the op-amp's input (+) pin to about +1.6V, just below the threshold required to light the red LED. Depending on the diodes' current/voltage characteristics, this usually requires 3 diodes. Put them in the circuit and use a multimeter to test the voltage drop between op-amp (+) and V–. Add a diode to raise the drop or remove one to decrease it.

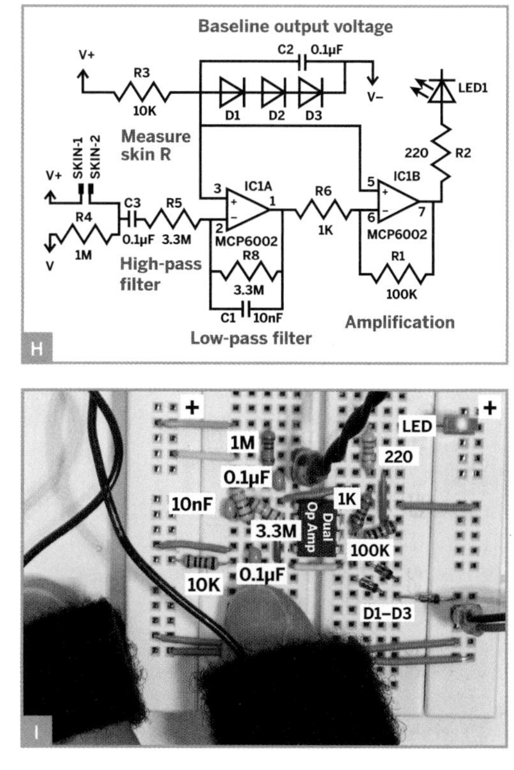

2e. For final amplification, so that the output voltage will cross the LED's threshold with signal spikes, a 100kΩ resistor sets the op-amp's amplification very high. If the circuit is too sensitive to GSR and the LED flashes too often, swap the resistor between the amp's (–) input and output for a 10kΩ–20kΩ, or use a potentiometer if you want more control.

3. TEST YOUR TRUTH METER

Insert your battery leads into the side rails of the breadboard (check orientation, red = V+), slip on the sensor cuffs (Figure G), and behold your very own Truth Meter.

See what happens when someone asks you questions or when you laugh or get surprised. Note the response has a 1- to 2-second delay.

Everyone responds differently. See if you can turn the LED on with your mind. Try it on your friends, acquaintances, or adversaries. It's a great way to get to know someone!

Photograph by Sean Montgomery

BRAIN BLINKER

J

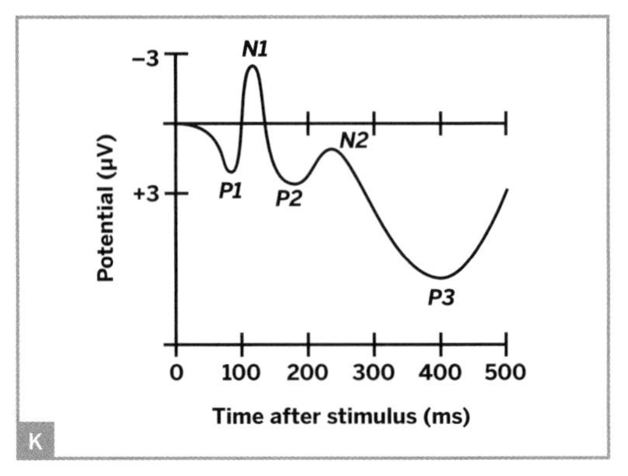

K

NEURONS IN CONCERT

Inside the brain, neurons communicate with one another using chemical and electrical signals. Each time a neuron fires, it creates a small voltage, and because neurons firing together communicate better than neurons firing out of sync, the brain uses systems of synchrony that fire specific large groups of neurons simultaneously.

This neural synchrony creates combined voltages large enough to be measured by an electroencephalogram (EEG), using electrodes placed on the scalp (Figure J). One example is the event-related potential (ERP) detected in response to an auditory stimulus (Figure K).

In addition to brief events like ERPs, the brain also exhibits rhythmically synchronized neuron firing over sustained periods, with different frequency ranges characterizing different activities.

During concentration, beta (14Hz–30Hz) brain waves appear in the front of the brain, whereas if you close your eyes and relax, alpha (8Hz–12Hz) oscillations appear at the rear. Theta (4Hz–7Hz) oscillations accompany drowsiness, and delta (1Hz–3Hz) accompany deep sleep. Gamma (30Hz–100Hz) oscillations are especially interesting to researchers because of their specificity; particular behaviors cause synchronized gamma levels to light up in different combinations of very specific regions.

A major difficulty of EEG is that the voltages on the scalp from neuronal activity can be just 1 microvolt or smaller (that's 1 millionth of a volt), which makes them 100+ times smaller than those generated by muscle activity in the eyes, face, head, and neck. Ambient electrical noise from wall power and other sources can also interfere with measurement of these miniscule voltages.

EEG machines used to cost tens of thousands of dollars, but today there are several good options for hobbyists. This project uses the NeuroSky MindSet, a $200 headset with many advantages: it's easy to use; it gives access to raw data and 10 interpreted measures; it uses open standards; it transmits its signal wirelessly (Bluetooth), which is safer than a wired connection; and it has dry electrodes, which means you don't need saline or gel. The MindSet's principal limitation is that it has only a single channel (on the forehead), which limits its ability to simultaneously monitor multiple brain regions.

Here's how to read brain-wave data from a MindSet into your computer for display and logging, and into an Arduino microcontroller for direct mind control of your physical device projects.

We'll begin by getting your MindSet plotting and logging data directly to your computer. For this, all you need is a NeuroSky MindSet and your computer. Then we'll make the MindSet talk to the Arduino microcontroller to light up an LED bar in response to your brain waves.

Photograph by Cody Pickens

1. SEE BRAIN WAVES (EEG) ON A COMPUTER

1a. Follow the instructions that came with your MindSet. First you'll install the software provided on the CD. Then you'll plug the included Bluetooth module into your computer, turn on the MindSet, and hold its play button to put it into Discovery mode (its red and blue lights will flash).

Open the Bluetooth settings on your computer and add a new Bluetooth device in custom mode. Choose the MindSet device and add a connection to the serial port. Note for later the serial port number that your MindSet connects to (e.g., COM41).

Finally, connect to the MindSet using the passkey 0000. You can test whether your MindSet is working properly by launching the Brainwave Visualizer included with the NeuroSky software.

MATERIALS AND TOOLS

Wireless EEG headset MindSet from NeuroSky (store.neurosky.com), $199. NeuroSky also recently released an Arduino port for its XWave iPhone headset (plxwave.com), $99.
Arduino Uno microcontroller part #MKSP4 from Maker Shed (makershed.com), $35, or older Diecimila version
BlueSMiRF Gold Bluetooth modem part #WRL-00582 from SparkFun Electronics (sparkfun.com), $65
Solderless breadboard RadioShack #276-001
Jumper wire kit RadioShack #276-173
LED bar, 10 segment SparkFun #COM-09936 or similar
LED, red
Resistors, ¼W, around 75Ω (11)
Computer with internet connection
Scissors, wire cutters, and strippers
Soldering iron and solder
Multimeter

1b. Once your MindSet is talking via Bluetooth to your computer, you can set it up to view and log the data in the Processing language using some great open source libraries. Follow the instructions at makezine.com/26/primer to download and install Processing, the libraries, and the MindSetBTViewer application.

1c. Edit *MindSetBTViewer.pde* to set the `serialPort` string variable to your serial port from Step 1a and to specify the variables you want to display in the `plotVars` string array. Be sure to include `Raw` and `ErrorRate` in `plotVars`; watching these values will help you adjust the headset later.

1d. Launch MindSetBTViewer and make sure it's receiving signal properly from the MindSet. Adjust the position of the device on your head to get a high-quality signal (this requires some patience). Look for low background noise in the Raw EEG and low Error Rates.

Many times you'll get a good signal very quickly, but other times it might take 30 seconds or more for the MindSet to settle into one. To speed the process, it often helps to dab your forehead and ears with salt water (mix ¼ teaspoon of salt into a cup of water).

If you get an error when you hit Run, check that you entered the correct serial port, your Bluetooth connection is set up properly, and your Processing libraries are in the correct location.

1e. Observe your brain waves and free your mind! Figure L (following page) shows examples of Raw EEG, Alpha 2 (10Hz–11.75Hz), Meditation, and Error Rate data from the MindSet. The top panel shows a high-quality recording; note the very low background noise in the Raw EEG (top trace) that permits both eye blinks and brain-wave patterns to be observed. We found that when you close your eyes, Alpha 2 and NeuroSky's proprietary Meditation values go up.

The bottom panels show examples of bad signal. On the left, the Error Rate is low and a Meditation value is consistently observed, but the background noise in the Raw EEG is too high to reliably observe brain patterns. This is typical in the first few seconds after putting on the MindSet. On the right, the Raw EEG is even noiser and error rates are high, making it impossible to observe anything.

2. CONTROL HARDWARE WITH YOUR MIND

Now let's read your brain waves into an Arduino to directly control hardware with your mind! Doing this involves a pair of Arduino programs: MindSetArduinoReader, which uses MindSet input to light LEDs while also writing the data out to a CSV (comma-separated values) serial data stream, and MindsetArduinoViewer, which reads live from the CSV serial stream and plots its data.

2a. Turn off your MindSet, then use jumper wires to wire up your breadboard with an Arduino, BlueSMiRF, and LEDs, as shown in Figures M and N, with the Arduino's digital I/O pins running to a 10-segment bar LED.

2b. Follow the instructions at makezine.com/26/primer to program your BlueSMiRF module with the specific MAC (media access control) address of your MindSet, using the Arduino sketch *CleanProgramBlueSMiRF.pde.* You'll open the CleanProgramBlueSMiRF sketch in the Arduino IDE and set your Board and Serial Port under the Tools menu. Then, change the MAC address (`mac[13]`) so that it reads your MindSet's MAC address, which can be found labeled on the MindSet or in your Bluetooth settings. Upload the program to the BlueSMiRF.

CleanProgramBlueSMiRF should start the single LED blinking on Arduino I/O 13 (Figure M, small red component, left side of breadboard). If the program fails, I/O 13 will not flash and the bar LED will light up to indicate where in the program sequence it failed. If this happens, try programming a second time.

2c. Disconnect the RX pin on the BlueSMiRF (TX on Arduino) to enter Run mode (Figure O). Download the MindSetArduinoReader folder from makezine.com/26/primer, extract it, and open *MindSetArduinoReader.pde* in the Arduino IDE. With your Arduino cabled to your computer, verify and upload the sketch.

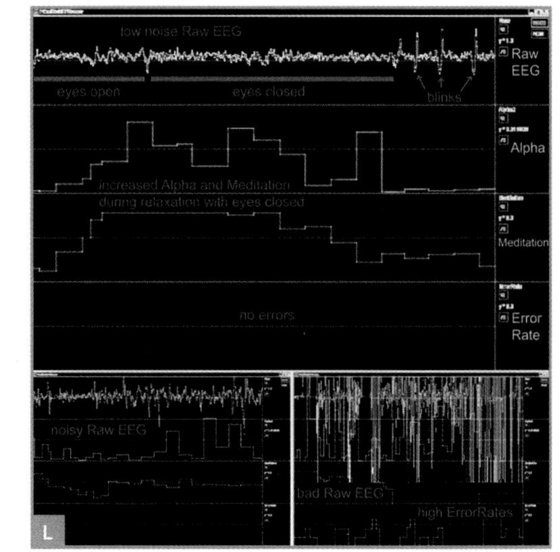

L

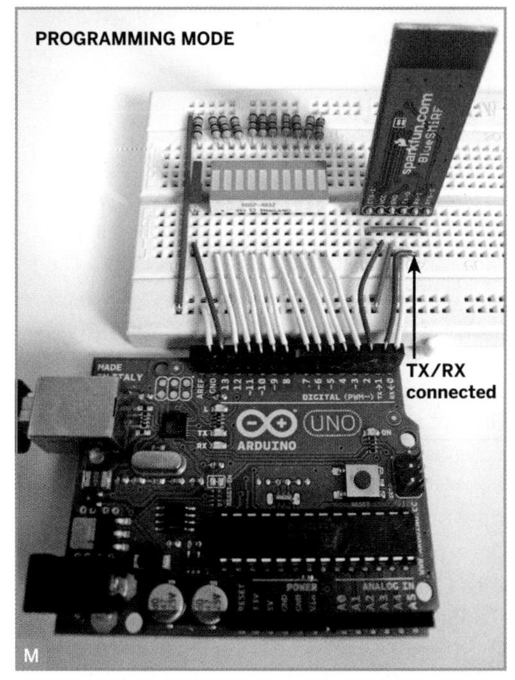

M

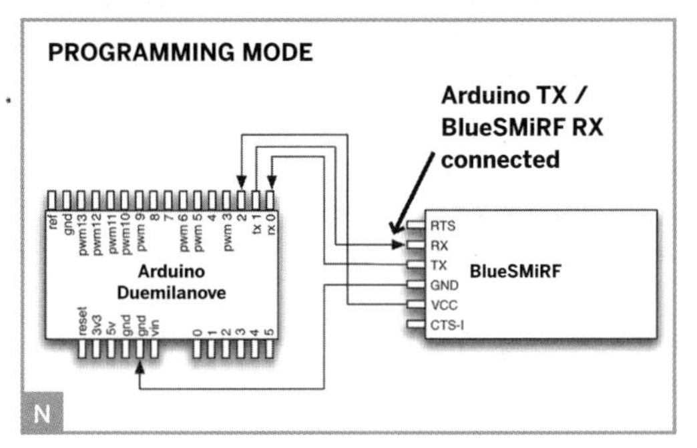

N

Photograph by Sean Montgomery

2d. Turn on your MindSet. After a short negotiation period, it should auto-connect to the BlueSMiRF module, indicated by a solid green LED on the BlueSMiRF and a solid blue LED on the MindSet. Open the Arduino IDE's Serial Monitor (under Tools). If everything is working properly, after a brief delay you should see several columns of CSV data streaming over the serial port.

2e. Close the Serial Monitor, then download and extract the MindSetArduinoViewer folder from the same URL. Open *MindSetArduinoViewer.pde* in Processing, change the `serialPort` definition at the top to that of your Arduino, and hit Run. This should open a window similar to the MindSetBTViewer showing Raw EEG at the top and other measures below.

2f. Put the MindSet on your head and try to get a good EEG signal, as in Step 1d. Once you do, your Error Rates should drop to zero, causing the red LED on your breadboard to light up. Meanwhile, the bar LED should begin to show your Meditation score — the more LEDs, the higher (Figure P).

You can change which values are sent over the serial port and which one displays on the bar LED in the "add your code here" portion of *MindSetArduinoReader.pde*. Make sure to change the `plotVar` labels in *MindSetArduinoViewer.pde* to correspond to any changes, so that your data is labeled correctly.

HACK THE DOORS OF PERCEPTION

Now you see how easy it is to begin biosensing. With the myriad possibilities for exploring the human body, we hope DIYers will discover many new ways to better understand themselves and communicate with each other and with the digital systems all around us.

BIOSENSOR RESOURCES

- » Kevin Kelly's Quantified Self resource links: makezine.com/go/quantified
- » *Honest Signals: How They Shape Our World* by Alex (Sandy) Pentland: makezine.com/go/pentland

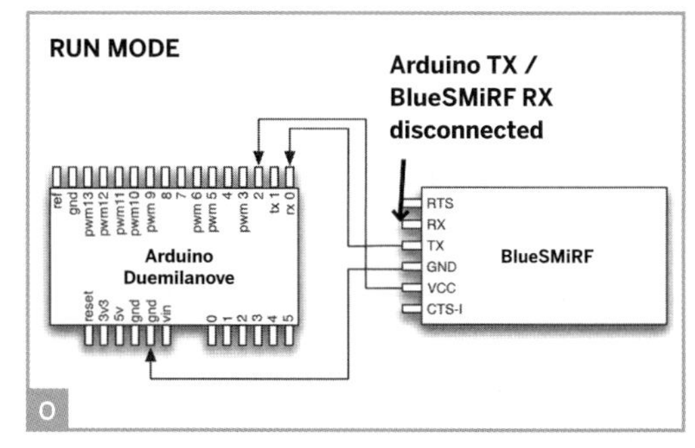

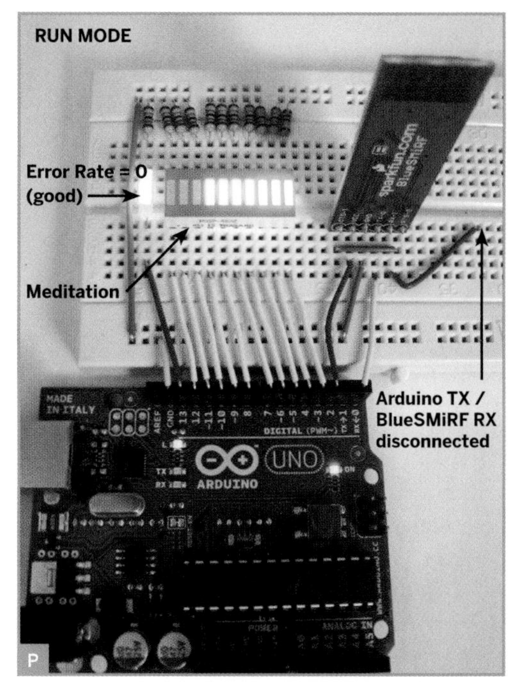

TIP: ***MindSetArduinoViewer.pde*** **can be used to plot** ***any*** **data sent over the Arduino serial port, not just MindSet data. Just change `plotVars` to reflect what you're sending, and voilà, you have a general-purpose digital oscilloscope at your disposal!**

SPECIAL THANKS

We are grateful to several individuals and organizations that contributed products, money, labor, and ideas to making this article possible. We are especially thankful to: NeuroSky, Inc., Nathan Seidle/SparkFun, Digi-Key Corporation, Rob King, Brendan Shrader, Michael Hogan, Jack Zylkin, and Ira Clavner. See makezine.com/26/primer for full credits.

1+2+3 Jumper Wires

By Charles Platt

You can make it!

WHEN YOU'RE WIRING A CIRCUIT ON a breadboard, neatness counts. If you use color-coded jumper wires that fit precisely, you can see your circuit clearly and track down errors easily.

I've tried both types of pre-cut jumper wires: the long, flexible ones with little plugs at each end, and the solid ones that are cut in 1⁄10" increments to match the breadboard spacing. The flexible ones make it easy to build a circuit, but you end up with a rat's nest in which errors are very hard to find.

On the other hand, the solid ones are color coded by length, which I find unhelpful. I want all my positive connections to be red, for instance, regardless of how long they are.

Cutting my own jumpers seemed a hassle. It isn't easy to trim ¼" of insulation from each end of a teeny wire, and get the length exactly right. What to do?

Instead of cutting the wire first and then trimming the insulation, I trim the insulation and then cut the wire.

YOU WILL NEED

- Breadboard
- Solid-core wire, 24 gauge
- Wire strippers
- Wire cutters

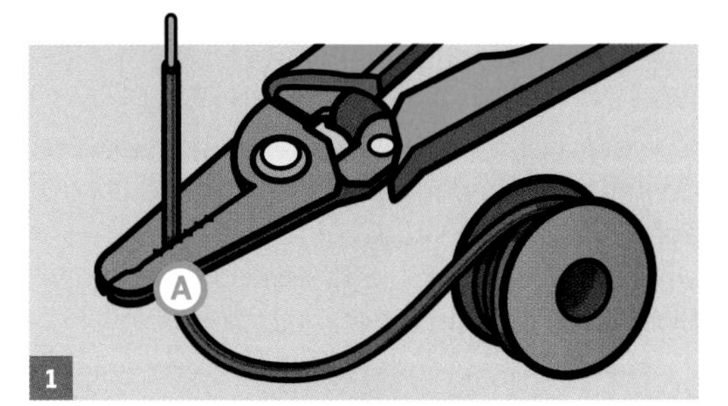

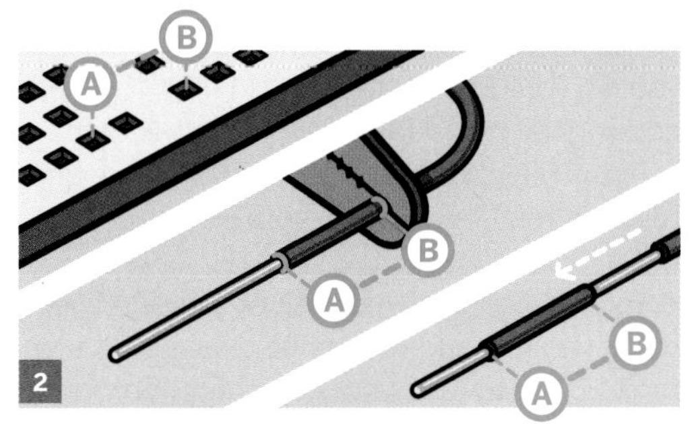

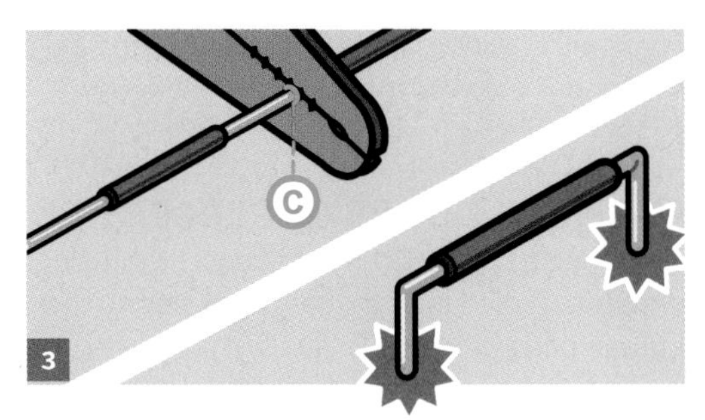

1. Strip the wire.

Unwind some 24-gauge solid-core wire from a spool, and remove a generous, arbitrary length of insulation by applying wire strippers around point A.

2. Trim and drag insulation.

On your breadboard, measure how long a jumper you need, then transfer this measurement to the insulation on your wire. In the diagram, it's the distance from A to B.

Apply wire strippers at point B and drag the insulation along until it's ¼" from the end of the bare wire.

3. Cut the wire.

Cut the wire at point C and bend the stripped ends, and your jumper should fit precisely.

Use It.

Now make more jumpers from the remaining length of stripped wire. If you create a variety in increments of 1⁄10" and sort them into a compartmentalized box, you can reuse them indefinitely.

Illustration by Damien Scogin

Charles Platt is a contributing editor of MAKE, and the author of *Make: Electronics*.

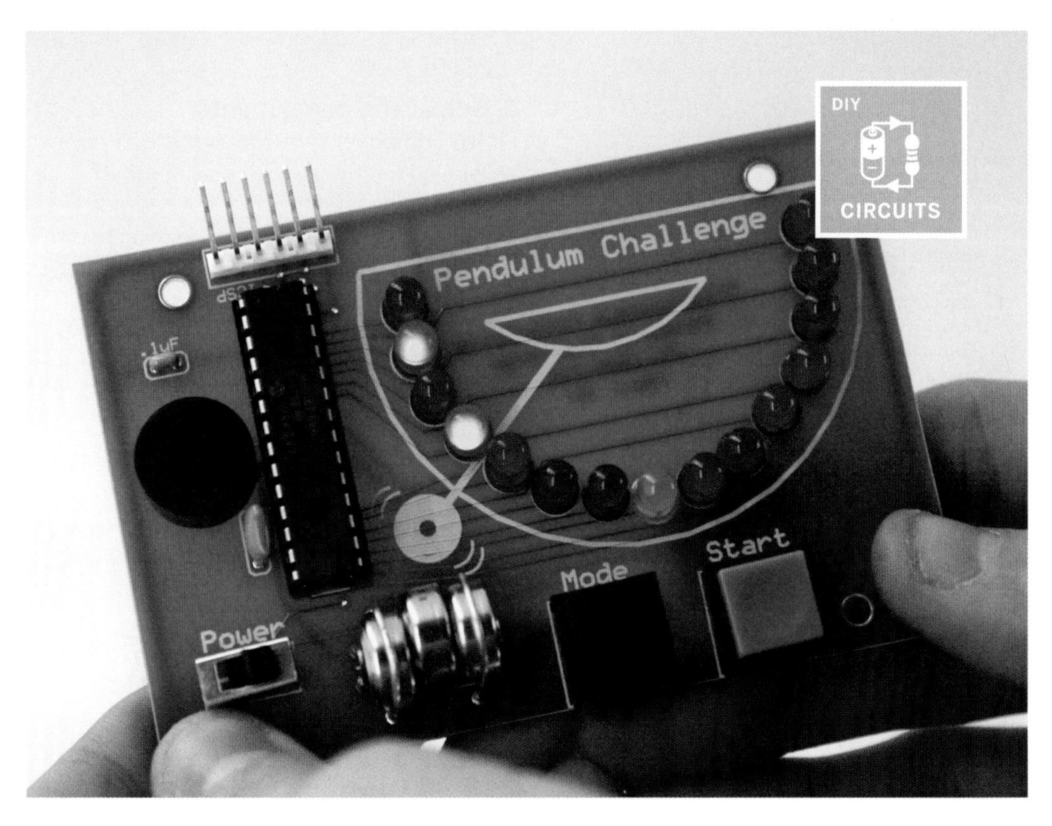

Pendulum Challenge

A fun reflex game for all ages.

By Ken Delahoussaye

I WAS RECENTLY ASKED TO SPEAK TO a classroom of college-bound high school students for a career day discussion. I gladly accepted the offer but soon wondered how I should present the profession of engineering to a generation that's already immersed in computers, cellphones, and MP3 players.

Realizing that my audience needed more than just talk, I decided to include the design and demonstration of an actual device. Then the idea hit me: a battery-operated, interactive gadget based on a game in a local arcade. The Pendulum Challenge was born.

Photograph by Sam Murphy

Game Description

The Pendulum Challenge is a handheld, single-player game consisting of a printed circuit board (PCB), batteries, and pushbutton switches. At the heart of the system, a PIC microcontroller runs the internal state machine that processes the game's switch inputs and drives its LED outputs.

The game's most distinct feature is its array of 15 LEDs (14 red and 1 green), arranged in the shape of an arc to simulate the path of a swinging pendulum. For sound effects, the game has a piezo buzzer.

The objective of the Pendulum Challenge is to stop the LED "pendulum" at the very bottom of its arc, on the single green LED. The green button starts or stops the virtual pendulum. Meanwhile, a black mode button toggles the buzzer so you can play in silence (a feature I added later for the benefit of hard-working parents and studious siblings).

The game progresses through 5 levels of play difficulty. The pendulum's speed increases through the first 3 levels, while levels 4 and

MATERIALS

Pendulum Challenge Kit product #MSPK01 from the Maker Shed (makershed.com/pendulum) has all the parts needed to make the game, including a pre-programmed PIC microcontroller. Or you can build the author's version by using the parts listed below:

Pendulum Challenge printed circuit board (PCB) Download the layout from makezine.com/26/pendulum and order through free ExpressPCB software (expresspcb.com). You can select the Mini-Board package (3 boards for $51 plus shipping).
PIC microcontroller part #PIC16F886-I/SP-ND from Digi-Key (digikey.com)
Resonator, ceramic Digi-Key #X909-ND
Piezo buzzer Digi-Key #102-1115-ND
Switch, SPDT slide Digi-Key #450-1598-ND
Switch, SPST pushbutton, green Digi-Key #EG2551-ND
Switch, SPST pushbutton, black Digi-Key #EG1301-ND
LED, green Digi-Key #67-1100-ND
LEDs, red (14) Digi-Key #67-1103-ND
IC socket, 28-pin Digi-Key #AE10286-ND
Capacitor, 0.1μF ceramic Digi-Key #BC1148CT-ND
Male header, 6-pin, right-angle Digi-Key #WM4104-ND
Coin batteries, AG13 or A76 (3)
Paper clips (2)
Panel or enclosure at least 3.8"×2.5", to mount PCB

TOOLS

Soldering iron and solder, pliers, wire cutters
Foam tape, double-sided (or hot glue)
Programming cable for PIC chips such as the PICkit 2, Digi-Key #PG164120-ND, $35
Computer with internet connection and Windows OS

5 incorporate random delays and reversals to keep you on your toes. At each level, the player is allowed 15 attempts to hit the target. If the target is hit, the player advances to the next level — if not, the game ends.

1. Get a circuit board.

The Pendulum Challenge is built on a custom 2-layer PCB, which you can make yourself or have manufactured. To design the PCB, I used the free Windows app ExpressPCB (expresspcb.com). You can download the PCB layout in ExpressPCB format at makezine.com/26/pendulum, along with the Pendulum Challenge circuit schematic and PIC software.

From ExpressPCB, you can easily price and order boards from the Layout menu. You'll typically receive the bare boards 3–4 days after ordering.

2. Solder the components.

Fit the components into their through-holes as labeled on the board and solder them in place, making sure to orient the cathode (–) side of all the LEDs (indicated with a shorter leg and a flat part on the plastic) through the holes with the square solder pads.

For the 6-pin programming header at the top, labeled ICSP (in-circuit serial programming), the best practice is to solder it in. But for aesthetics, I sometimes leave it unsoldered. Then I just temporarily plug the header into its holes whenever I need to program the chip, and extract it afterward. I can't guarantee that this will always work, but it has for me so far.

3. Make the battery holder.

I couldn't find a suitable battery holder to fit the available space on the board, so I constructed my own using paper clips. (Optionally, you could use an external AA battery holder, soldering the black and red wire leads to appropriate pads on the PCB.) My paper-clip fabrication is composed of 4 parts: 2 retainers and 2 terminals.

The retainers are U-shaped wires that hold the batteries in place; to make them, bend paper-clip wire around a coin cell, run each end through one of the PCB hole pairs above and below the battery site, and cut to length for soldering.

For the terminals, bend 2 shorter V-shaped wire pieces and insert them into the holes at each end of the batteries. Solder all 4 parts onto the circuit board (Figure A).

4. Install the software.

Install the software that comes with your PICkit programmer, then download the Pendulum Challenge software from makezine.com/26/pendulum. You can grab either the compiled hex image *PC1000.hex*, which is ready to upload to the PIC, or the source code directory if you want to read or modify and compile the code yourself using PIC development tools.

Make sure the power on the PCB is

switched off, then plug your PICkit 2 (or other) programming cable between your computer's USB and the PCB's 6-pin header (Figure B). The PICkit software should automatically detect the presence of a PICkit device. Load *PC1000.hex* onto the chip by navigating to its folder location and clicking the Load button. Disconnect the programming cable, and the Pendulum Challenge is ready to play.

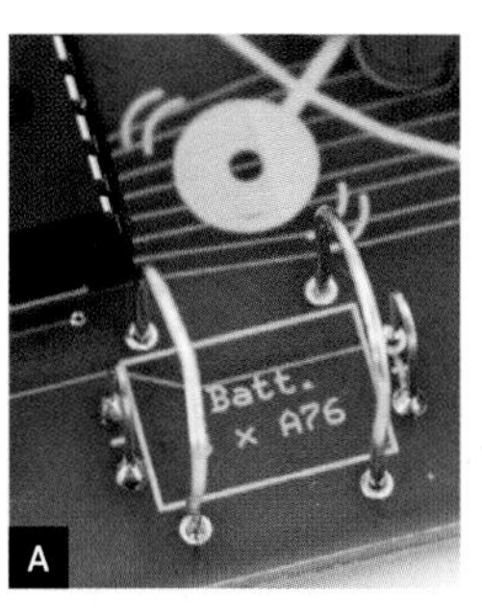

A

⚠ WARNINGS: **Don't solder the batteries themselves, as this may result in explosion and release of toxic fumes. Before connecting the programming cable, always make sure the PCB's power switch is in the off position, and don't turn the power back on until the programmer has been disconnected.**

B

5. Choose a panel or enclosure.

To frame and carry the PCB and prevent short circuits on its soldered side, I mounted it to an aluminum panel using double-sided foam tape strips (hot glue could work too if it's thick enough). The panel had previously been the lid for an enclosure, and I drilled a hole in each corner to accommodate 4 plastic hole plugs as protective feet (Figure C). You can mount the PCB to just about any flat object it will stick to or mount it inside an enclosure.

6. Let the fun begin.

Carefully insert batteries if you haven't already, observing the proper polarities. Slide the power switch on (to the right) and push the green Start button to initiate the virtual pendulum. Push the green button again to attempt to stop the pendulum on the target. To enable silent mode, hold the black button down for 2 seconds.

As mentioned, you'll have up to 15 tries total, after which the game shows your score by illuminating the corresponding number of LEDs. As with golf, lower is better.

You'll find that the game is quite intuitive, and in no time you'll be challenging your friends to see who can attain the best score.

The Pendulum Challenge was a real hit at career day. Everyone wanted to try it, and it made several trips around the room as students wanted another chance to improve their score. I was pleased with this reaction. And ever since, the Pendulum Challenge has been getting frequent workouts from my 12-year-old daughter, who still enjoys the challenge and continues to better her score.

C

Download the Pendulum Challenge PCB layout, circuit schematic, and software at makezine.com/26/pendulum. Get the complete kit at makershed.com/pendulum.

Ken Delahoussaye (kdelahou@att.net) is a software engineer and consultant in Melbourne, Fla., who specializes in embedded firmware and PC applications. He also maintains Kadtronix (kadtronix.com), a website featuring automation and control resources.

Photography by Ken Delahoussaye, Sam Murphy (Figure A)

Hack Electronic Pushbuttons

Tap into your electronic devices and take control.

By Peter Edwards

IN THIS TUTORIAL I'LL EXPLAIN HOW you can easily hack the controls of almost any electronic device. Why would you want to? Maybe you want to rewire a Nintendo joystick to your computer so you can control Mario via Max/MSP. Maybe you want to set up magnetic sensors to steer a remote control car while you're tap-dancing. Or wire the buttons in your TV remote to big, arcade-style buttons mounted in your coffee table. Or modify a musical instrument (as I'll show here). There are countless possibilities.

This guide applies to most button-hacking projects but there are always exceptions to the norm and baffling anomalies. These techniques will work for many circuits but not all.

What's a Button?

A pushbutton is a simple electromechanical device that makes or breaks an electrical connection when activated. Hold 2 pieces of wire in your hands. Connect the ends together, now disconnect them. You just performed the functions of a pushbutton.

There are many varieties of pushbuttons but the most common (and simplest) is the *momentary SPST (single-pole, single-throw) switch*. This will often be listed as "(NO)" which stands for "normally open." The parentheses around NO tell you it's a momentary switch — the circuit stays closed only while you keep your finger on the button.

This button has 2 connection terminals or nodes. Activating (or pushing) the button connects these terminals together. That's it! This sends a signal to the circuit telling it to do something specific. It doesn't matter how the nodes are connected; all the circuit knows

Photograph by Sam Murphy

KINDS OF BUTTONS

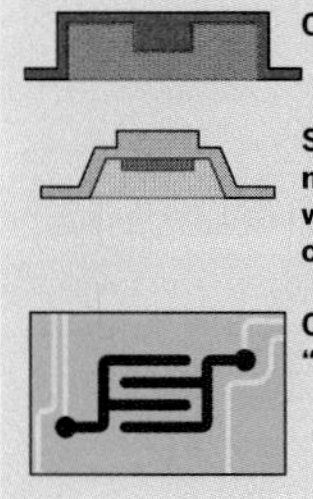

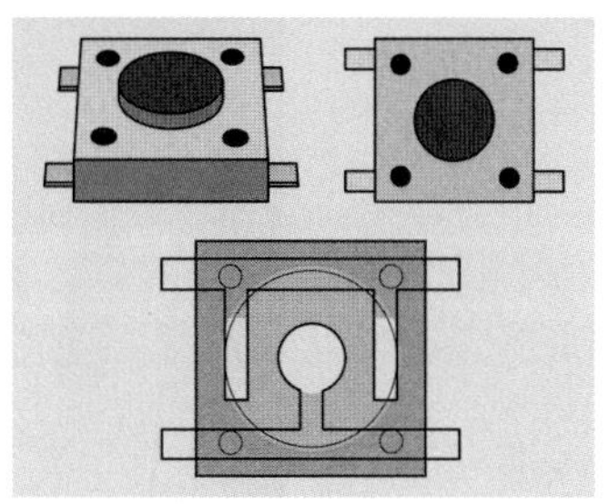

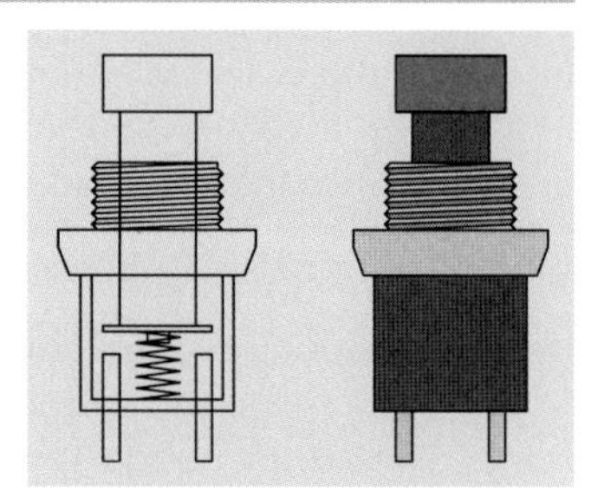

CONDUCTIVE RUBBER BUTTON

The cheapest and most common style of button used in consumer electronics. It's easy to install and can be made in a wide range of shapes. When you press the button, it pushes the conductive rubber membrane against 2 interlinked sets of "fingers," which are printed on the board in conductive ink.

Pros: Cheap and reliable.

Cons: This style of button requires special fabrication techniques that are unavailable to hobbyists. It's also the most difficult style to hack, and because it's made of conductive rubber it can handle only very low current.

TACTILE BUTTON

Used where conductive rubber isn't feasible, or in higher-current applications. This button is cheap, easy to use, and easy to hack. A tactile switch has 4 legs; the legs across from each other are typically connected internally.

The only drawback for hobbyists is that it must be board-mounted, which presents challenges over panel-mount pushbuttons.

Pros: Cheap and reliable, easy to hack, low resistance, can pass higher current than rubber switches.

Cons: Small, must be mounted on a circuit board.

PANEL-MOUNT BUTTON

Rarely found in consumer electronics but a favorite with hobbyists. These switches are the most versatile but also the most expensive and, relative to tactile and rubber switches, the most delicate.

Pros: Easy to use, available in many different configurations, can handle high current, very easy to hack.

Cons: Much more expensive than other kinds of buttons. More moving parts, therefore more delicate.

⚠ CAUTION: As always, only work with battery-powered electronics or circuits that are powered with a wall wart adapter. Don't tinker with circuits that plug directly into the wall unless you know what you're doing and are qualified to work with deadly voltage levels.

is that they are. That means anywhere there's a button, you can replace it with another button, a sensor, a relay, or any other means of passing and breaking current flow. As long as the nodes connect and disconnect, it'll work.

Just make sure the button you use can handle the electrical current that will pass through it. This article only covers low-current circuits and buttons.

Let's Hack Some Buttons

I like making music, so I decided to hack the buttons in a bunch of musical toys so that I could control them all with a sequencer.
I selected 3 Casio keyboards — one for the drumbeat and 2 for the melody — and 3 voice memo recorders to sample and play back the sound from the keyboards, introducing all kinds of interesting variables into the music.

MATERIALS AND TOOLS

Soldering iron, solder, and insulated wire
NPN transistors general purpose
Resistors, 10kΩ (optional) and 100kΩ
Arduino microcontroller Maker Shed part #MKSP4, makershed.com, $35
Breadboard
Multimeter with continuity tester
Mini screwdrivers for opening electronic devices

1. Identify the buttons you want to hack.

Let's look at the Casio SA-38. This keyboard is useful because it has 5 big drum buttons (Figure A, following page). I can hack these and sequence them to make drumbeats.

2. Open it up.

Once you open up the keyboard, you'll see 3 circuit boards (Figure B). The green board handles power and audio. Underneath it is a brown board with a big chip on it. This chip is the main brain of the keyboard, so I'll call this the control board. This board and the third board alongside it hold all the conductive rubber pushbuttons. The control board has the

Illustration by Peter Edwards

function buttons, including the drum buttons I want to get at (shown by the red box). The third board has the keyboard buttons.

3. Find your buttons.

Unscrew the control board and turn it over. Now you can see all the button contact points. The 5 drum button contact points each have 2 nodes (Figure C). These nodes look like 2 hands with their fingers interlaced.

4. Connect your wires.

The contact nodes are printed on the board in conductive ink, which is impossible to solder to. Follow the leads trailing away from each node and you can see that one node of each button is connected to a solder point nearby.

You can also see that the bottom nodes of the 3 leftmost buttons are connected together, and so are the 2 right buttons. This is a common trick used in digital circuits to trigger several functions with just a few pins. There's no obvious solder point near these nodes, so use your continuity tester to find their 2 solder points elsewhere on the board.

Solder a wire to each node's solder point, then test it by touching the ends of your wires together to trigger each sound. Finally, write down the button configuration. I use colored ribbon cable to make this easy (Figure D).

5. Hack more buttons!

The Casio SK1 keyboard lets you record sequences then play them back one note at a time by pressing One Key Play. I hacked this button so I can control playback (Figure E).

On the voice recorders, I hacked the Record and Play Back buttons (Figure F). These samplers are easy to mod and cheap to buy from All Electronics (makezine.com/go/recorder).

6. Choose switches to interface with.

To replace a conductive rubber switch, you can use any of the following:

High-Voltage/Low-Resistance Switches

» Panel-mount and tactile pushbuttons
» Magnetic reed switches
» Relays

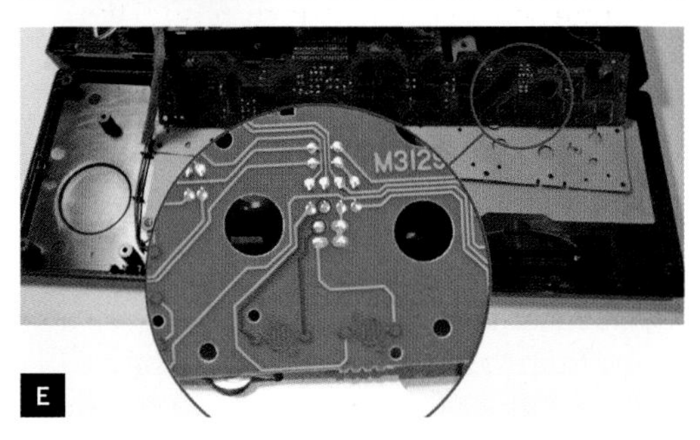

Photography and diagrams by Peter Edwards

Low-Voltage/High-Resistance Switches

» Transistors (My favorite! See Step 7.)

» Switching ICs (such as the CD4016)

Panel-mount pushbuttons and relays are necessary for low-resistance, mid-voltage (more than 1V) applications such as passing audio signals or powering circuitry.

For the sequencer shown in Figure G, I used relays as my switching mechanism to allow for the greatest variety of application. Plans for building your own analog step sequencer are available at makezine.com/go/sequencer.

7. Connect hacked buttons to Arduino.

I use an Arduino microcontroller and some additional transistors to trigger the buttons I hacked. This is a very flexible method that I also use to interface devices with my modular synthesizer.

Transistors are small, cheap, and amazingly powerful. I use general purpose NPN style, 2N3904 or 2N2222 (Figures H and I). I'm sure lots of other NPN transistors will work.

Wire up the switch as shown in Figure I, and test it on a breadboard before soldering. This configuration will work in most cases, but experimentation may be necessary. If your hacked device is triggering erratically, install a 10kΩ resistor to ground. If it's still acting up, try increasing the value of the 100kΩ resistor. If it's not triggering, reduce the value of the 100kΩ. If it still doesn't work, try different transistors. There's a combination out there that will work!

Also, you should connect the digital output pins to the base pins of your transistors through 100kΩ resistors. When the pin goes HIGH it will trigger the hacked button.

I wrote some simple code to control the Casio keyboards and samplers, and recorded some videos and music, which you can check out at: makezine.com/go/sequencercode. Plus, stay tuned for my tutorial on building sequencers in an upcoming issue of MAKE.

Peter Edwards is a circuit-bending and creative-electronics pioneer in Troy, N.Y. He builds electronic musical instruments for a living at Casper Electronics (casperelectronics.com).

NOTE: You must connect the ground point on the circuit you're hacking to ground of the Arduino. If you're using battery-powered electronics, just connect the negative battery terminal to the GND pin on the Arduino. If you don't do this, it won't work.

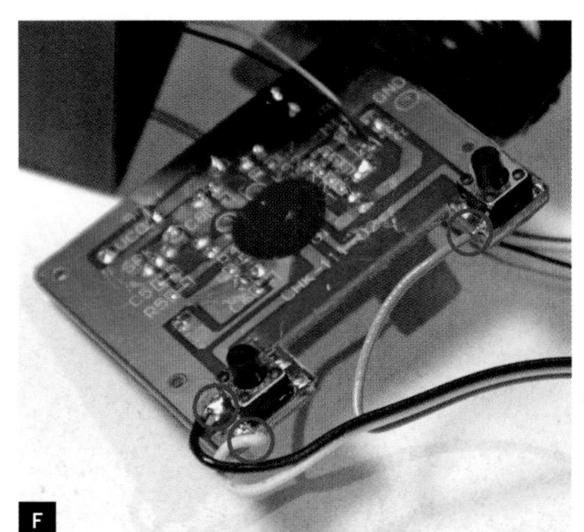

F

G

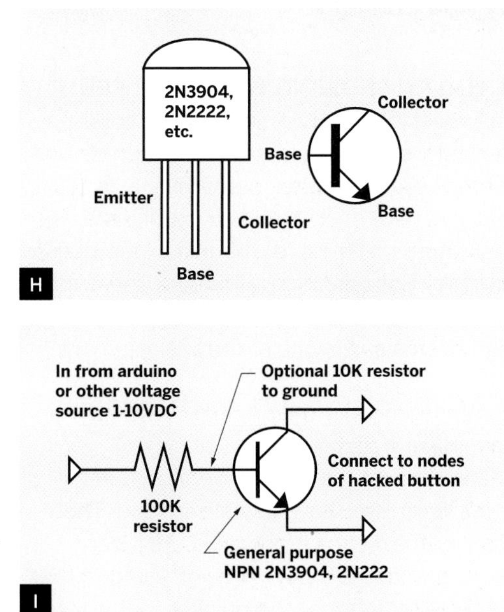

H

I

Solar Food Dryer

Drying with the sun is one of the easiest, most energy-efficient ways to preserve your harvest.

By Abe Connally

IF YOU'RE GROWING YOUR OWN FRUIT and vegetables, or just trying to eat food that's produced locally, you come across one obvious problem: when something is in season, you have more than you can handle, and then there's nothing for the rest of the year.

The obvious solution is to preserve your food when you have it in abundance, and drying with the sun is one of the easiest, most energy-efficient ways to do so. It also maintains a lot more of the original nutrients than canning or freezing.

The concept of a solar dryer is simple: move warm air over thinly sliced food. The warmer the air, the more moisture it can remove from the food. However, you don't want the air to move too quickly, as it will cause the temperature to decrease. Our design creates just enough air movement and warmth to dry food quickly.

The food is laid out on trays, which sit behind a transparent polycarbonate sheet. Below the trays there's a metal shelf, painted black, that serves as a heat absorber. As heated air rises through the food, cool air is drawn in through the bottom vent, and the heated, moisture-laden air flows out the exhaust at the top.

Because the dryer is something we plan to use for many years to come, we decided to make ours out of metal. If you don't have access to a welder, you can make the frame out of wood, but you'll need to adjust these plans accordingly.

Photography by Josie Moores

MATERIALS

Steel tubing, 1" square, 40'
Sheet metal, galvanized, 16'×3'
Polycarbonate panel, flat, 2'×8', translucent sold for greenhouses and patio covers
Hinges (2) and latch
Silicone caulk
Dimensional lumber, 1×2, 8' lengths (11)
Food-safe screen, 16'×2'
Thin wooden molding, 48" lengths (2)
Sheet metal screws, self-tapping (100 or so)
Wood screws (50 or so)
Paint, black

TOOLS

Welder, metal chop saw, drill, tinsnips, tape measure and marker, framing square, wood saw, box cutter, staple gun, staples

A

1. Make the frame.

You can make the frame any size you want, but I settled on 48" long by 18" wide. This was the size that I could cover with one sheet of 24"×96" polycarbonate.

Cut 3 lengths of square tubing 46" long, and 2 lengths of 48". These will be the horizontal beams of the frame.

Cut 2 lengths 34" long and 2 lengths of 22", for the upright pieces of the frame.

Place the two 34" uprights on a level surface. Mark each one 3" from the bottom and 10" from the top. Place a 46" beam between the 2 uprights, below the 3" mark, and weld it in place. Place another beam between the 2 uprights, above the 10" mark, and weld it in place. This is the front (or door) panel.

Weld the 22" uprights between the 48" beams to form a big rectangle. Be sure to put the beams above and below the uprights, not between them, so that everything will come out to be 48" long. This is the back panel.

With the back and front panels welded, it's time to lay out the sides. Stand the back panel up vertically on a level surface. For the bottom sidepieces, cut two 20" lengths of square tubing and lay them down, butted against the back panel, one on each side. Stand the front panel up vertically on top of these sidepieces. (Clamp or tie both vertical panels to something to keep them upright.) Square the bottom sidepieces with the uprights and clamp them all in place.

Now cut 2 lengths of square tubing that will connect the tops of the front and back panels. Because the front is taller than the back, these 2 top sidepieces will be angled. Hold a piece of square tubing roughly in place and mark it so that it fits in between the front and back panel. Repeat for the other side. If your panels are level and square, these top sidepieces should be the same size.

Clamp the 2 top sidepieces in place between the front and back panels, and then weld the 4 clamped joints, checking square periodically.

Lie the unit down, so that the taller, front (door) panel is on the floor and the bottom is facing you. Insert the last 46" horizontal beam into the bottom panel, about 4" up from the front panel. This will frame the bottom vent. The dryer frame is done (Figure A).

2. Make the door.

Cut two 47" beams and two 19½" uprights, and place them over the opening of the front (door) panel to check for size. Make sure you put the beams above and below the uprights, not in between them, so that the door will measure 47" long by 21½" tall. You want the door panel to overlap the frame about ¼" on all sides. Weld the door frame together.

Cut a piece of sheet metal 47"×21" and attach it to the door frame using the drill and sheet metal screws. The door is done. You'll attach it after everything else is completed.

3. Add tray supports and the heat absorber.

Before you cover the frame, install the tray supports while there's still plenty of room to work. Start by cutting ten 21" lengths of 1×2 lumber.

Mark all 4 of the uprights 4" up from the bottom bar. Proceed to mark the uprights at 4" intervals. You should have 5 marks on each upright.

Using sheet metal screws, attach the 1×2 boards inside the dryer frame, between the front and back panels at each mark (Figure B).

Cut a piece of sheet metal 48"×18" for the heat absorber. The absorber sheet sits on top of the lowest tray support boards and the bottom beam of the door panel. (You'll need to cut out its front corners to fit around the uprights.) Screw it in place with wood screws and sheet metal screws. The absorber sheet doesn't go all the way to the back panel; there's a gap of about 3½".

4. Cover the frame.

Cut the polycarbonate sheet in half, to give you two 24"×48" sheets. Use a box cutter to score and then slice the polycarbonate.

Clamp each piece to the outside of your welded frame, one on top, and one on the back panel. Make sure they cover the frame well. Don't clamp so tightly that the polycarbonate breaks.

Pre-drill holes around the perimeter of the polycarbonate for screws, and use sheet metal screws to attach the polycarbonate to the metal frame (Figure C).

Cut pieces of galvanized sheet metal to cover the sides. It's easiest if you cut a sheet 21" wide by 35" long and then trace the frame out on the metal to get an exact fit. Screw the sheets to the frame with sheet metal screws.

Also cut a piece of sheet metal 18"×48" to cover the bottom, but don't attach it yet.

There will be a 4" gap on the front (door) panel side of the bottom plate. This gap will become the intake vent.

Cut pieces of sheet metal 4"×48" and 8"×48" but don't attach them yet. You'll attach these to the front panel, below and above the door opening, respectively, leaving a 2" gap at the very top for the exhaust vent.

5. Screen the vents.

Cut 2 pieces of screen, 6"×50", which will cover the intake and exhaust vents to prevent insects from entering the dryer.

Cut 2 lengths of molding, 48" and 46". Staple one of the pieces of screen to the 48" molding, along the edge of the screen. Using a few small screws, attach this molding to the underside of the top polycarbonate panel, so that the screen hangs down the front (door) panel. Attach the sides of this screen to the dryer frame using one screw on each side (Figure D).

Place the other piece of molding between the uprights of the front (door) panel on the inside of the sheet metal at the top. Screw it in place and staple the loose edge of the screen to it.

Attach the bottom screen with a few screws on the sidebars to hold it in place (Figure E).

Now screw the sheet metal to the bottom and front (door) panel. Make sure the screen lies in between the sheet metal and frame for a secure and insect-tight fit (Figure F).

6. Seal and paint.

Caulk all the edges and seams of the sheet metal and polycarbonate panels with silicone caulk. Paint the entire inside black, focusing especially on the heat absorber and sides.

7. Make the drying trays.

Cut 1×2 lumber to 8 lengths of 46" and 12 lengths of 19½". Screw and glue these boards together into 4 rectangular tray frames measuring 46"×21½" with a support board in the middle of each. Make sure the rectangles are nice and square.

Cut 4 pieces of screen 54"×28". Fold screen around the edges of each frame and staple it to the boards, pulling it tight as you go along.

Attach 2 screws for guides on the underside of each tray, one on each side. The screws should be about 1" from the end of the tray. Don't screw them in all the way; leave about ½" of the screw sticking out (Figure G). These guides make it easier to slide the tray in and out of the dryer.

8. Attach the door.

Attach the door with hinges and a latch (Figure H). And you're done!

Use It

Before using, make sure the dryer sits in the sun for a few days to allow any fumes from the paint and silicone to escape. Then test the dryer using 2 trays at a time, increasing to 4 trays if the weather is clear and dry. Slice food as thin as possible (¼") to hasten drying time.

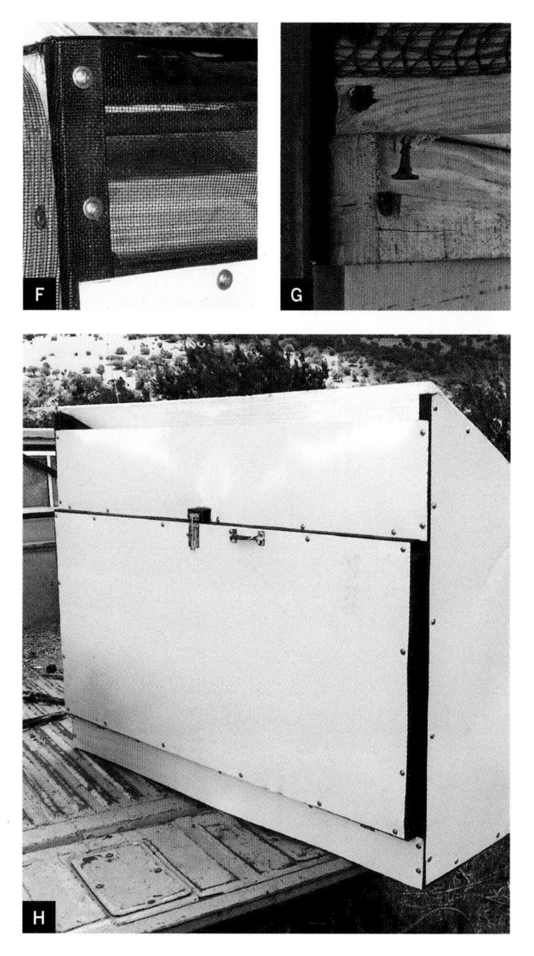

I like to keep an oven thermometer in the dryer to see the temperature inside. Ideally, you want it to be about 130°F inside. If the temperature is lower than this, reduce the size of the bottom vent with a piece of cloth.

Foods can be rehydrated before using, if needed. You can dry leaves, roots, fruits, vegetables, nuts, fruit pulp, and anything else you can think of.

More photos and food drying resources at: velacreations.com/solarfooddryer.html

Abe Connally and Josie Moores are an adventurous couple living in an off-grid hideaway with their 2-year-old. Their experiments with energy, architecture, and sustainable systems are documented at velacreations.com.

Fabric Formed Posts

Create cheap and easy reinforced concrete columns in organic, sculptural forms.

By Abe Connally

FABRIC FORMING IS ONE OF THE EASIEST and cheapest ways to form concrete. Instead of using time-consuming and costly wooden or metal forms, you use little more than vinyl, plastic, or similar flexible fabric.

The end result is not only aesthetically beautiful, but it can also be stronger because it lends itself to rounded shapes, compacted pours, and consistent curing.

The basic principle is simple. Cut your fabric to the required size, attach it to solid stays along both sides, which are then screwed together and braced to hold the form in place, and pour the concrete into the flexible form (Figure A).

This method can be used for many shapes and creations, but I'll focus on a simple stand-alone column. You can employ a similar method (as I'll note) to make posts that attach to an existing wall, which is a little easier.

Count on about 2 hours for digging the posthole and pouring the base, then another 2–3 hours later on for 2 people to make the post itself. Part of creating the post involves making the fabric form, which will cost roughly $8 (or around $25 if you have to purchase the 1×4s) and can be reused. The materials for each post will be around $25.

1. Pour the base.

Dig a hole about 12" in diameter where you'd like your post to stand. The depth should be ¼ of the post's height for a fairly lightweight

Photography by Josie Moores

load or 1⁄3 of its height for something very heavy, like an earth-retaining wall. Our 8' posts support a lightweight acrylic concrete roof, so we dug the holes about 24" deep.

Make the bottom of the hole bell-shaped (wider at the very base). This makes the post harder to pull up in an extreme wind.

Place your rebar column or other metal reinforcement in the hole, raised up on rocks, plastic chunks, or any other solid and non-decaying junk you have around. The key is that you want concrete under any metal, so there's no chance it will rust. If the rebar touches the earth, it could wick water and slowly dissolve.

Level the metal reinforcement to stand upright, making sure you level it both east-west and north-south. Brace the column in place by leaning a couple of long scrap boards or sturdy pieces of metal against it (Figure B, following page).

If your post will be right next to or touching an existing wall, you can brace it by attaching it to the wall itself, like with C-clamps, using blocks as standoffs to create a gap in between. The gap should be half the diameter of the final post minus half the width of the rebar column.

Mix a batch of concrete in the wheelbarrow (see the Note at top right), adding just enough water until it's barely workable. It should be just soupy enough to fill the hole, but don't make it any soupier than necessary — the drier concrete starts out, the stronger it will be.

Use a scoop to pour concrete into the hole to just above ground level. Tamp it down every so often to ensure that it fills the hole completely.

Let the base set for a couple of days before pouring the column. Once the concrete has set, you can remove the braces from the metal column, and the base will hold it in place.

2. Make the form.

The size of your form will depend on the size of your post. The dimensions described here are for a post that's 8' tall and 8" in diameter. Adjust these dimensions according to your needs.

Cut a piece of tarp 8' long × 36" wide. On a flat surface, lay two 8' 1×4s flat and parallel, 22" apart, or a little bit less than you want your post's circumference to be, because the tarp will stretch a little. (If the post will be touching

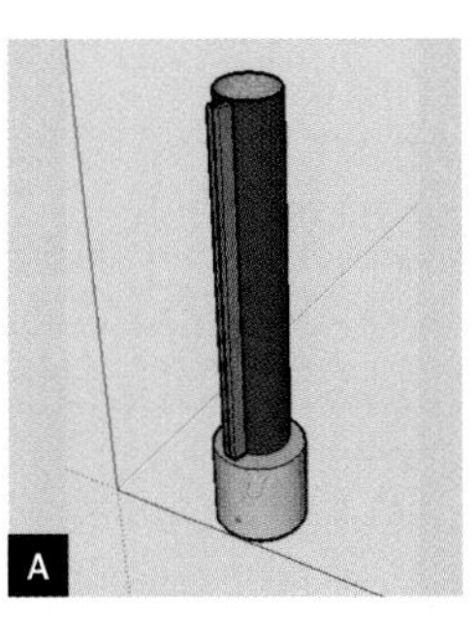

NOTE: To calculate the volume of your base to determine the necessary amount of concrete, use the formula $\pi r^2 h$, where r is radius and h is height. You'll want to make sure to calculate the radius based on the bottom of the bell-shaped hole so you'll have enough concrete.

MATERIALS

Rebar cage column, length determined by the height of the post plus the depth of the base We bought prefab column consisting of four ¼" rebar (steel reinforcement bar) uprights with ⅛" horizontal stays welded around at regular intervals. If you can't find the pre-welded version, you can lash together three ⅜" rebar uprights with a thick wire stay, wrapping the wire around each rod as it spirals along the column's length. Or, instead of rebar, you could use remesh (welded wire mesh) cut and bent into a long cylinder.

Concrete (cement, sand, gravel, water), a bit more volume of dry ingredients than would fill the column and base dimensions We use a 1:3 ratio of cement to a sand/gravel mix (in our case, gravelly sand dug from a nearby river), with about a 1:2 gravel/sand ratio. There are lots of varieties of cement, sand, and gravel (and even a commercial product called Quickcrete, which is premixed sand/gravel/cement). Use the cheapest and most basic.

Vinyl, plastic, or fabric tarp, large enough to wrap column Vinyl is best, since it stretches much less.

Lumber, 1×4 (2) length equal to the post height

Staples, a lot about 2 dozen per foot of post height

Wood screws, about 1½" long (3–4)

TOOLS

Tape measure

Level

Staple gun

Post-hole digger or rock bar

Long wooden boards or sturdy pieces of metal (3–4) for bracing

Drill and drill bit sized to drill pilots for wood screws

Concrete mixing tools: wheelbarrow, shovel, scoop, and measuring cans We use 2 old coffee cans — one for dry ingredients, and one for water.

Rocks, bricks, sandbags, or similar enough to arrange around the base of the fabric form to contain the wet concrete an existing wall, place the boards closer together to subtract the amount of the post's perimeter that will be touching the wall).

Lay the tarp flat on top of the 1×4s. There should be an extra 7" or so of tarp overlapping the 1×4s on both sides. Fold this back in on itself on each side, and staple the tarp to the wood close to the inside edge (Figure C). Put a staple every inch or two — the concrete will exert quite a force on the tarp, so more is better than less.

As you staple the other side of the tarp to the second 1×4, periodically check the distance between the 2 boards (22" in this case) to make sure it stays uniform (Figure D).

Align the 1×4s with stapled sides together, sandwiching the tarp's edges. Drill pilot holes through both boards so they can screw together flat against each other — top, bottom, and one or two in the middle.

3. Pour the column.

If you haven't already, remove the braces that held the rebar column in place. Stand the form up, and wrap the fabric around the rebar so that the stapled sides of the two 1×4s meet flat against each other. Having 2 people helps with this. Screw the boards together, and level them upright along 2 different planes. Brace the form boards in position using long boards or metal pieces, as before. (If your post is against a wall, just secure each form board against the wall on either side of the rebar.)

You can use the formula given earlier ($\pi r^2 h$) to calculate the amount of concrete needed for your post. To avoid concrete working itself out at the bottom of the fabric, stack some rocks, sandbags, etc., around the base. This should prevent much from coming out down there.

Pour concrete into the fabric, patting around the outside of the form as you go to help compact the concrete and release air bubbles that can make the post weaker. When the fabric becomes taut, you've patted enough — if you pat too much, you risk stretching it. When my wife, Josie, and I do this, she usually stays at the top of the ladder,

PURO YONKE TIPS **If you use a sturdy, thick plastic and keep it out of the sun during storage, these forms can be reused several times. Old tarps or awnings are good sources of fabric. Look for woven fabrics with interesting textures and strengths. Stapled areas are notorious for stretching, so double- or triple-fold the fabric there to help make it last longer. Using nails with washers to secure the fabric in place might be better.**

B

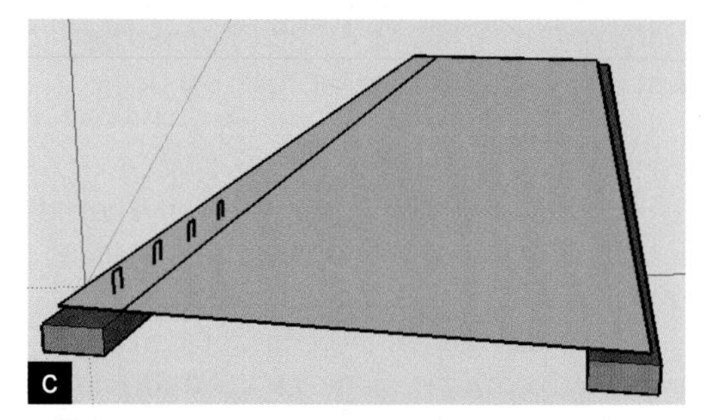

C

D

and I pass her coffee cans of concrete. She pours it into the form, and I do the patting to make sure everything settles.

You don't have to pour the entire post at once, but if you don't, the post will have a seam and won't be as strong. For an 8' by 8" post, we can pour it all in 15–20 minutes, and it takes 2–3 wheelbarrows of concrete to fill up the form (Figure E). Pouring goes fast; you can empty a wheelbarrow of concrete in less time than it usually takes to mix it, even if you're mixing by hand.

When the concrete is hard to the touch, take down the braces, unscrew the screws, and carefully peel off the form (Figure F). You can store it for reuse. And although vinyl stretches less, we've found that concrete sets far quicker in a plastic form than in vinyl — we don't know why.

E

F

Get Creative

There are all kinds of things you can do to a fabric form to make the end result more interesting. Small holes are not a problem with holding wet concrete, so you can use many types of fabric, and any texture, wrinkles, or tie lines on the tarp will be transferred to the concrete. For example, if you wrap rope across the tarp when it's empty, the concrete will bulge out around the string, leaving a woven pattern in it (Figures G and H).

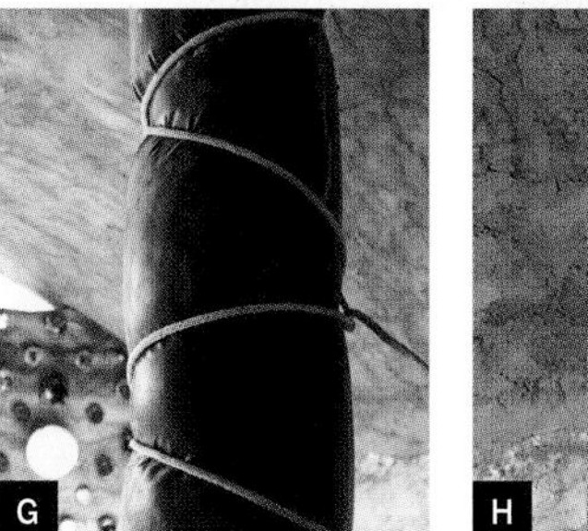
G

H

Resources

- Abe and Josie's fabric formed posts page: velacreations.com/fabricforms.html
- Centre for Architectural Structures and Technology (CAST, University of Manitoba), Fabric Formwork: umanitoba.ca/cast_building/research/fabric_formwork
- Fab-Form, fabric-formed concrete supplies: fab-form.com

Fool's Stool

Build a fake Colonial Period stool good enough to fool almost everybody.

By Gordon Thorburn

ALMOST ANYBODY CAN MAKE A 17th-century board stool. In those early, pioneering times, techniques and tools were fairly primitive and ambitions consequently modest, so today, faking the American Colonial style requires a do-it-for-fun, cavalier attitude rather than serious precision.

In making your brand-new antique, it actually helps if you begin with joinery that's not perfectly straight, level, smooth, and right-angled.

You're not making a direct copy. The photographs are here for guidance, but each fake piece is an individual article. It's not so much a clone as a close relation of something genuine.

When British settlers turned up in America, they brought the Jacobean style with them and translated it as American Colonial, but they couldn't go to the wood merchant and buy planed and sanded oak, cut for them in convenient widths and lengths. They didn't have an electric jigsaw or a power drill, but you probably do. You'll also need a hammer and a few odds and ends. Compared to the Pilgrim Fathers, you have it very easy.

1. Cut the 5 pieces of wood.

We're not going to give precise dimensions, because they don't matter. You have 5 pieces to cut: a top, 2 identical legs, and 2 identical sides. Draw templates on paper, and pencil in the outlines on the wood (Figure A).

The height of the stool is about the same as the length of the top. The top is between 18" and 20" long, with grain running longways,

Photograph by Sam Murphy

MATERIALS AND TOOLS

Oak boards, 18"–20" long: 12"–14"×¾" (1), 10"–12"×½" (2), and 8"×½" (2)
Pine scrap, about 1"×1"×4"
Paste finishing wax, and wood glue
Cow manure (preferred) or wood stain and shellac
Drill, jigsaw, handsaw, circular saw, hammer, chisel, spokeshave
Brush (optional)

12" to 14" wide, and ¾" thick.

Legs are about 2" narrower than the top, to give the overhang. Sides are about 8" wide. Decorative cuttings-out with your jigsaw must be kept simple. You could just drill a few large holes in a pattern.

The 4 pieces of the frame are joined by 4"-long, ½"-wide slots (Figure B). To obtain the squared-off end on the slots, drill a hole in each corner of the slots so the jigsaw blade can make the 90° angle (Figures C and D).

The slots in the sides should be at a slight angle. If the side is 18" long, the gap inside the legs at the top edge should be about 12" and at the bottom edge about 13½".

2. Distress the pieces.

The fate of most of your stool's rustic and very early brethren was to be chopped up and put on the fire when the more elegant stuff came in. For it still to be here in the 21st century, it must have suffered in many ways, to be rescued at last by your good self or another saintly person you know.

Before assembling the stool, rub with coarse sandpaper and throw and drag your cut pieces of oak around the yard or along the street until you get a satisfactory number of chips and scratches (Figure E).

3. Glue the pieces together.

The Colonial maker would have used nothing but joinery to hold the frame together, but then his stool was only going to be sat on. His stool wasn't going to be thrown around, soaked, shot at, and otherwise attacked (see Steps 5 and 7). You shall use wood glue (Figures F–H).

Depending on the accuracy of your slot

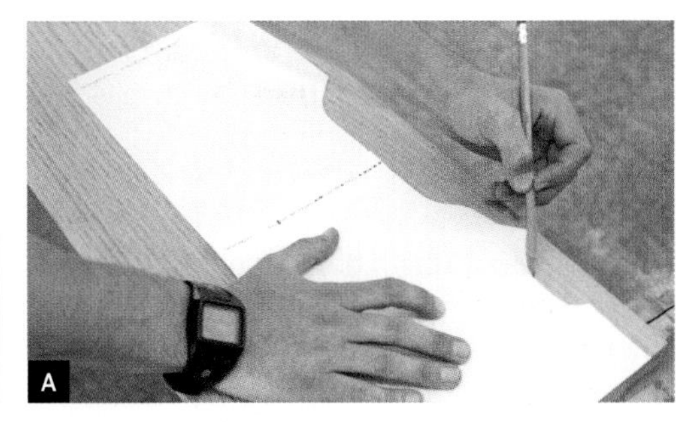
A

B

C

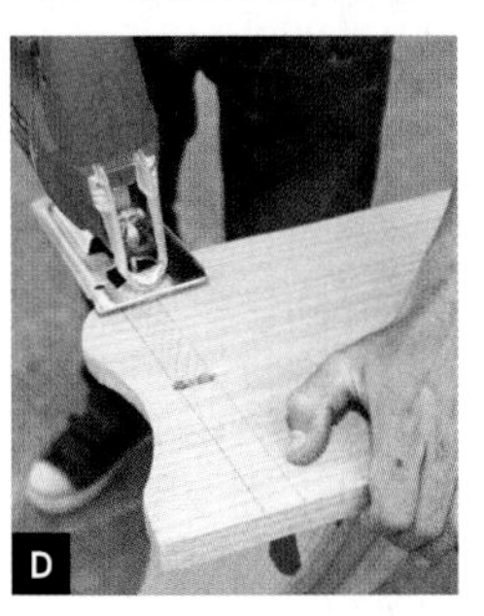
D

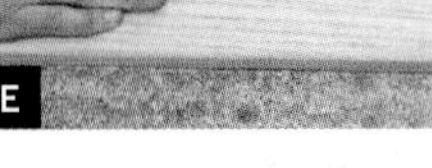
E

F

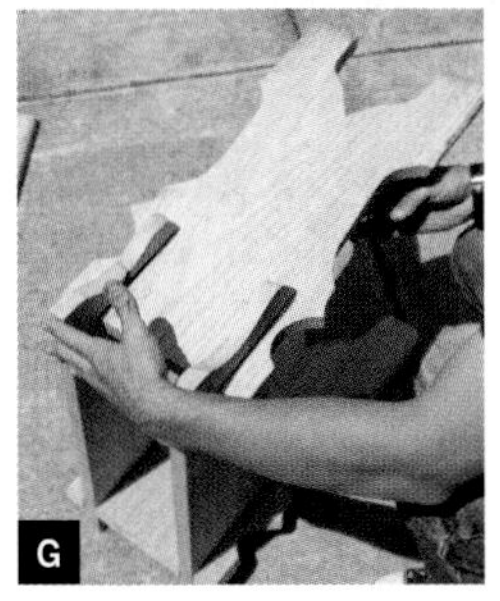
G

H

Photography by Ed Troxell

cutting, you may need to tie string around the leg pieces to keep them in order while the glue sets.

4. Fix the top.

Drill ¼" holes through the top into the legs, one at each corner, and 2 more at the mid-point into the sides (Figure I). Do this by eye. Don't measure.

Use a chisel and hammer to make 8 square pegs out of pine wood (Figures J and K). Use a piece of pine that has a nice straight grain to make it easier when splitting. Make square-section lengths, like very fat matchsticks, slightly thicker than the holes you've drilled. If your peg-wood has good straight grain you can split it with a chisel; otherwise saw it.

Drop some glue in the hole and bang the peg in fearlessly with a hammer (Figure L). Saw off any surplus, not too close (Figure M).

5. Distress the stool a bit more.

The expert looks for a texture to the wood on the seat, with the grain standing out a little, the pegs proud, and a general air of being used and worn and knocked about. Obvious places for wear are the bottoms of the legs and the edges of the seat. Also, the underside of the overhang should show the patina of being picked up a million times by greasy fingers.

So, how do we fool our friends and neighbors? First, add wear, again with coarse sandpaper and a spokeshave. Simulate dog bites with vise grips (Figure N). Don't overdo it, but allow no line to be perfectly straight.

6. Stain with cow muck.

Next, it's the color. With oak, there's only one way to get that perfect, 500-year-old look, and that is to make your stool and wait 500 years.

The best compromise between time and convenience is to steep your oak in cow muck. We realize this is not an option everyone can pursue, but then the best never is. The process gives you a very good color that's not flat, as stain tends to be, but variegated in a natural, haphazard way.

You could go and collect a sack full of cow

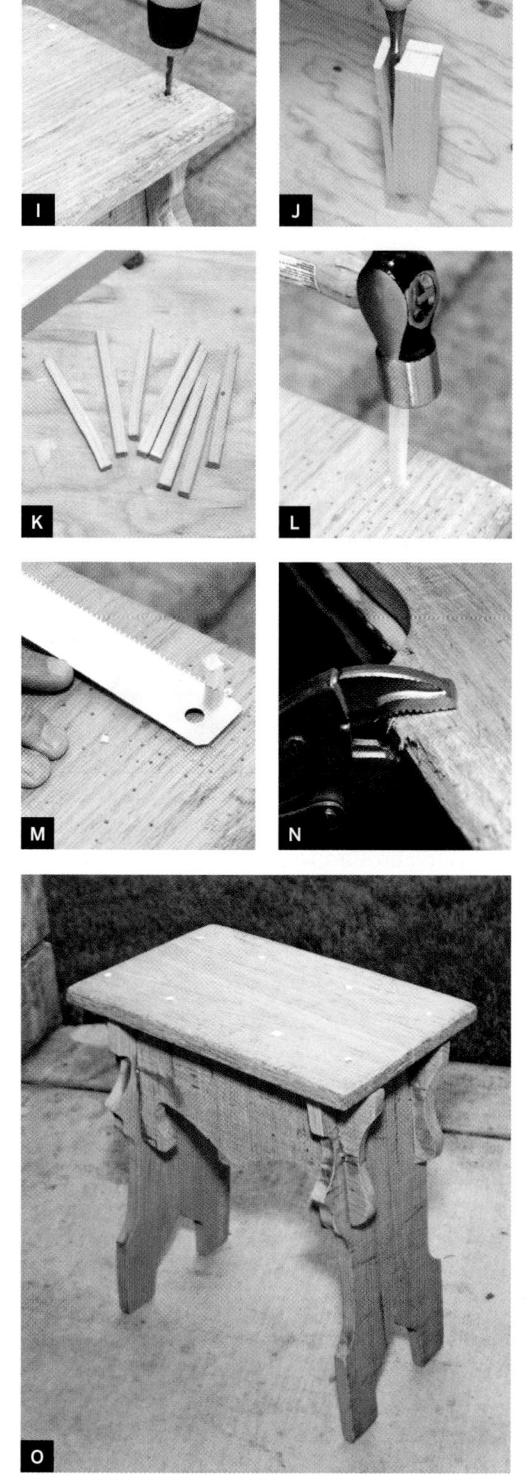

pats, make them into a soupy slop in a trash can, and immerse your stool in it. Add your used tea leaves and coffee grounds. Compost is good. Four to six weeks will be a reasonable time in the pit but a mid-term inspection will be necessary. The color revealed when wet is the color you'll get when it's polished.

Once you're satisfied with the color, you'll have to get rid of the smell. A week in running water will do it, so it's handy to have your own trout stream or access thereto, or a week in the sea. Otherwise it's a fortnight or more in a rain barrel with regular water changes, lots of wet weather, or the equivalent in garden hose soakings. When the wood is wet, that's the time to add wear to the bottoms of the legs. Tap with a hammer to splay the feet and round them off, to simulate years of scraping on a stone floor.

Drying out in sun and wind between soakings seems to help lose the smell. Don't be dismayed by the light gray it goes when dry. Polish will bring it all back. On no account use any chemicals or proprietary deodorizers. These may spoil your color and/or seal in the smell so it will never go away.

Another great thing about soaking is that you're likely to get the odd random crack or two (Figures P and Q), which adds an authenticity you could never get on purpose. It also raises the grain as if a hundred family backsides have gradually worn away the softer parts of the wood, leaving the harder grain standing out in tiny, well-polished ridges.

If cow muck isn't for you, you have alternatives, such as stain. This is the last resort because it tends to give a flat result and it's the hardest to get right.

If you have no other option, then at least do a week's soaking in water, bash the feet, and allow the stool to dry out thoroughly, preferably in the sun.

Mix dark oak stain with shellac, about 50:50 (try more or less to your liking). They don't blend, which is the idea. When you like the result, slap it all over the stool with a brush. Do this at speed, with confidence. Don't worry about runs and uneven areas.

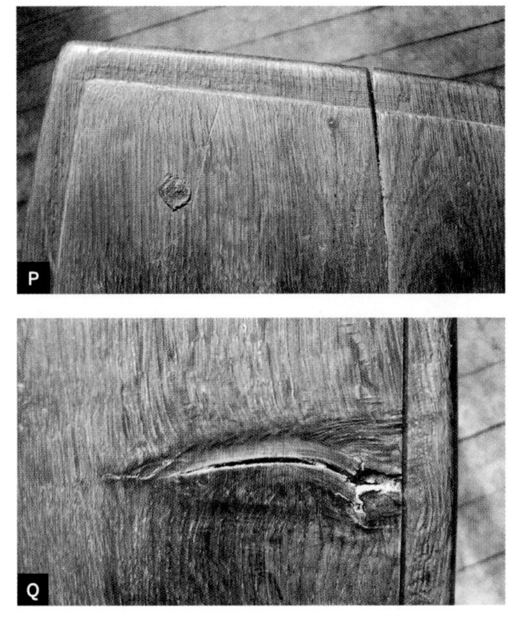

When it's dry, go into the shade and rub a little wax furniture polish on it. Phew. Let's hope it works.

7. Distress the stool even more.

We give here a menu of possibilities for mistreatment, but be circumspect. No one piece could have suffered all of these miniature disasters. Beware of overfaking.

Use it as a sawbench. Poke holes in it with an awl. Scratch it with a dinner fork. Carve initials on the top in a corner, or at the top of a leg. Draw a rudimentary coat-of-arms with a soldering iron. Knock off a vulnerable bit and glue it back on. Scorch part of it with a blowtorch. Stain it with sloe gin or ink. Burn it with a cigar or cigarette. Some of these techniques will reveal new wood beneath. Black shoe polish will sort that.

Now you can polish. Rub all over with diluted shellac to seal the color, then wax polish except for the main underside, which never saw the light and was never touched. And that's it. Best of luck. Go on, you can do it.

Gordon Thorburn is a British author of more than 20 books, most recently *Cassius*, the true story of an extraordinary police dog, and *No Need to Die*, about American volunteers in the RAF Bomber Command during World War 2. His most successful in the United States has been *Men and Sheds*.

Photography by Gordon Thorburn

Jigsaw Puzzle Chair

Make this smart chair from a 2'×3' sheet of plywood.

By Spike Carlsen

I CALL THIS A JIGSAW PUZZLE CHAIR FOR 3 reasons: the pieces are cut from plywood like a big jigsaw puzzle; you can cut it out using only a jigsaw; and when you tell people it's made from only a 2'×3' sheet of plywood, they'll be puzzled over how you could be so clever.

This chair is so clever, in fact, you can build 5 of them from a single 4'×8' plywood sheet.

1. Mark out the parts.

Use the measurements in the cut diagram (Figure A) to mark out the straight lines on a 2'×3' sheet of plywood. A framing square or drywall square will speed up the process. Make sure the chair sides are identical, and the seat and back are likewise.

Create the 1" radius on the corners using a compass or small jar (Figure B).

2. Cut out the parts.

You can make all the cuts with a jigsaw, but a circular saw — one with a fine-tooth, plywood-cutting blade to avoid splintered edges — will make the straight cuts faster (Figure C). See "Making the Plunge" for how to do this safely.

Sand the edges smooth.

3. Assemble the 5 parts.

Use clamps to hold the side frames upright, then position the seat so it overhangs each of the side frames by 1". Secure the seat to the side frames using 2" drywall screws.

Adjust your jigsaw or circular saw to cut at a 10° bevel, then square off the ends of the 4"×14" center scrap so it fits snugly between the front legs to help wiggle-proof the chair. Screw it in place (Figure D).

Finally, lay the chair flat, center the back on

Photography by Bill Zuehlke Photography

MATERIALS

Plywood, ¾"×2'×3' Less expensive plywood has a "good side" and a "bad side"; if you use it, you'll wind up with a chair that has one good side facing in and the other good side facing out — a scenario that would make Louis Vuitton roll over in his grave. We suggest you use plywood with 2 good sides.
Drywall screws, 2" (13 or more)
Optional: Wood glue, wood putty, primer, and paint

TOOLS

Jigsaw
Cordless drill with driver bit for drywall screws
Clamps (2)
Sandpaper or power sander
Optional: Framing square or drywall square, drawing compass or small jar, circular saw with fine-tooth plywood-cutting blade, router with roundover bit

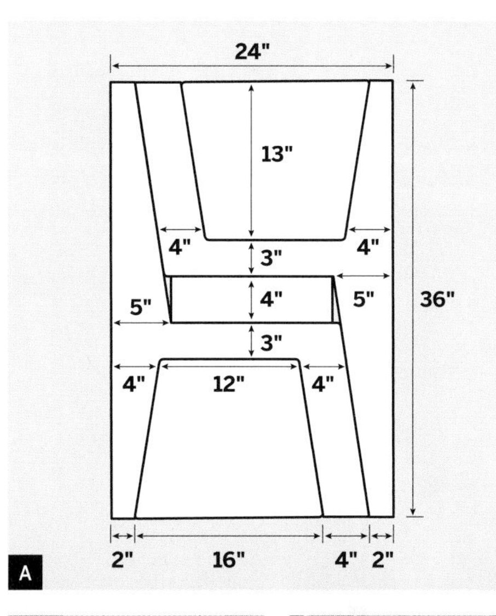

A

TIPS: If you own a router, use a "roundover" bit to soften the edges of the components before assembling the chairs.

If you're going to make several chairs, use one side frame that you've cut as a tracing pattern.

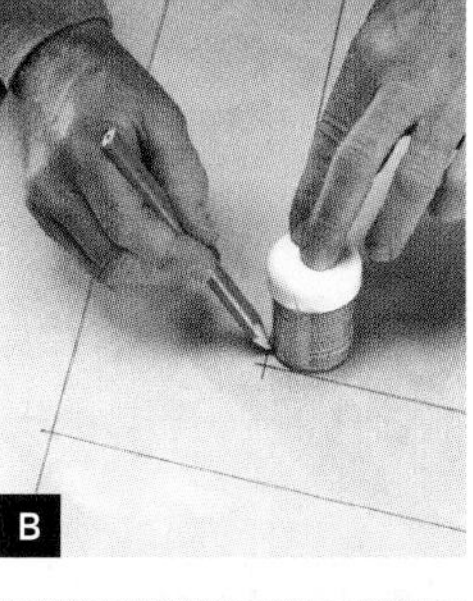
B

C

the uprights of the side frames, and screw the back into place.

4. Finish (optional).

Fill the screw holes with putty and let it dry. Sand the chair smooth, then apply a coat of primer and 2 coats of paint.

D

CAUTION: If you approach this timidly, the saw can catch on the plywood and kick back toward you. If you've never made a plunge cut, practice on scrap plywood until you get it right. If you still don't feel comfortable, use a jigsaw for all your cuts.

To see build videos of the Jigsaw Puzzle Chair and other easy furniture projects, visit ridiculouslysimplefurniture.com.

Excerpted from Spike Carlsen's new book *Ridiculously Simple Furniture Projects*, Linden Publishing, ISBN #978-1-610-350-0-44.

Spike Carlsen was the executive editor of *The Family Handyman* magazine, and is a regular contributor to *American Woodworker*, *FreshHome*, and *Men's Health*. He is the author of *A Splintered History of Wood: Belt Sander Races, Blind Woodworkers and Baseball Bats*.

MAKING THE PLUNGE

The secret to making accurate plunge cuts is to work decisively. Set the cutting depth of your circular saw to about ⅞". Align the blade with your cut line, place the nose of the saw firmly on the plywood, then retract the blade guard. Turn on the saw, then slowly — but firmly — lower the spinning blade into the plywood and push the saw forward.

As you lower the blade, the initial plunge cut will be several inches long. Position your saw along the line so the back of the blade doesn't cut where you don't want it to cut (Figure C).

Hidden-Magnet Knife Holder

Craft a cool way to keep your cutlery.

By Larry Cotton

ONE SUMMER MY FAMILY RENTED A beach cottage that had a special kitchen knife. We started calling it our go-to knife. Everything we did, from cutting melon to filleting a fish, we did with this knife.

It was such a nice knife that I eyeballed it closely and discovered it was a Fixwell (fixwell.com). Well, I'd never heard of a Fixwell knife, but it turns out it's the one that Subway restaurants use, so how could it not be good?

Long story short, I bought two sets of 12 knives (Sharp Tip style), made six 4-knife holders, and gave five away for Christmas presents. Friends and relatives have apparently found even more uses for these amazing knives than we did. So order some knives!

Fixwell knives are distributed in the United States by Alfi International (alfi.com), which has the broadest selection of them online. The knives cost under $2 (if you buy 12) up to about $5 (if you buy just one).

They're available with two slightly different handles and are extremely sharp (trust me on this one). Of course, you'll need a holder for your knives, and since Alfi doesn't make one (yet), here's how to make your own.

1. Collect the materials.

First get some neodymium magnets, approximately ½" in diameter and ⅛" thick. They're much stronger (and more costly) than the larger black ones commonly sold as refrigerator magnets. But you gotta get the more expensive ones for this knife holder — you don't want your holder dropping knives.

The only other material you'll need for this project is some wood: a piece of 1"-thick stuff, and some ¼" plywood. The 1"-thick wood

Photograph by Sam Murphy

MATERIALS

Small neodymium magnets (1 per knife) part #44204-5 (disc magnet D12.5mm×5mm) from Indigo Instruments (makezine.com/go/indigo)
Piece of wood, 1" thick, at least 2"×6" I prefer hardwoods, but any wood will do. Length and depth depend on how (and whether) you'll round the edges of the holder. See body text for more info.
Piece of appearance-grade ¼" plywood slightly longer than knife height by final width of knife holder
Wood screws, flathead (Phillips), #6, ½" long (2) to attach holder to backing board
Wood screws, flathead, #8, ¾" long (2) to attach backing board to cabinet door
Wax or stain
Flat file, approximately 8"–10" for a medium-smooth cut
7⁄16"-diameter wood dowel, about 6" long to press magnets into hole
Super glue
Painter's tape or hot glue

TOOLS

Drill, drill bits, and countersink bits to match the wood screws
Spade (½") or brad-point bit
Wood vise
Router, with a ½"-radius rounding-over bit
Sander or sandpaper
Band saw
Circular saw or table saw (optional) can cut wood stock to 1" thick, but a band saw is recommended
Hacksaw or fine-toothed saw (optional) can be used in place of band saw to cut grooves in holder

CAUTION: When using power tools, always wear eye protection, work slowly and carefully, and follow the operation and safety instructions in the owner's manual.

NOTE: The 1" thickness is suggested to achieve the desired look for your holder, but if you don't have access to (or aren't comfortable with) power tools, a standard 2×2 from the hardware store can be used (its actual dimensions are 1½"×1½").

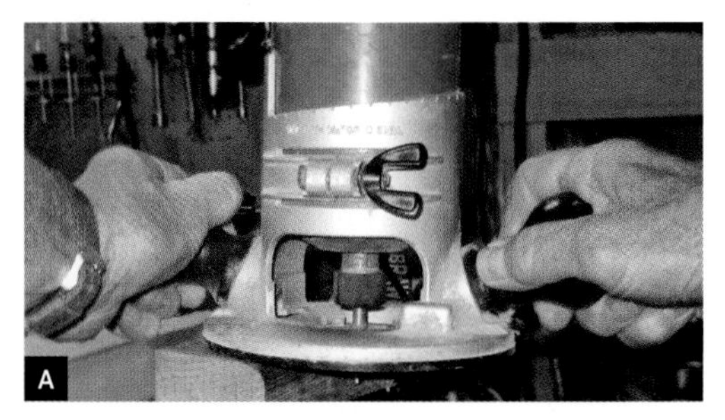

A

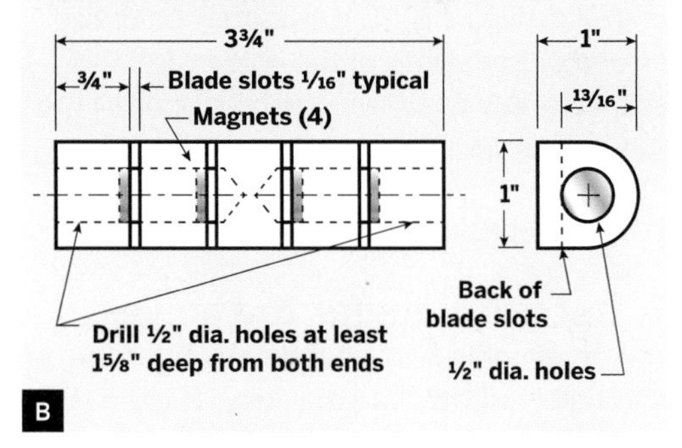

B

may be a bit difficult to find, but you can cut thicker wood to 1" thickness using a band saw (recommended). A table saw or circular saw can be used, but they don't offer nearly as much control. Most any wood will do, but go for something with a bit of interesting grain or a darker wood like walnut. I had some 1" birch and teak on hand, so I used that. The plywood should be a decent grade, with at least 5 plies. Mahogany, birch, maple, or oak look nice.

For safety, your 1"-thick wood needs to be wider and longer than the final piece to be cut from it. You'll be rounding its corners with a very sharp router bit revolving at high speed, so a larger piece of wood will help keep your hands away from the cutter. If you don't have a router, you can use a 1"×1" piece of wood and leave the corners square, but soften the front edges with a sander or sandpaper.

Photography and diagrams by Larry Cotton

2. Form the holder.

If using a handheld router, you'll want the wood dimensions to be at least 1"×2"×6" (longer and deeper is fine). Clamp the 1"-thick wood tightly in a vise and round 2 corners with a ½"-radius rounding-over bit in a router turned right side up (Figure A). Make the rounded corners at least 4" long.

If you have a larger piece of wood (at least 18" long and 2" deep), you can clamp the router and move the wood across the bit. Again, make the rounded corners at least 4" long.

Cut the piece (we can call it a holder now) to its final length, which depends on how many knives you want to hold. Since Alfi sells only 1, 3, 6, or 12 knives as sets, plan on 3 or 4 slots to hold your Fixwells. If you're going to hold 4 knives, the holder should be about 3¾" long; for a 3-knife holder, cut it 3" long (see

Figure B, previous page). Again, feel free to experiment with different dimensions.

Sand the holder to blend the 2 curved surfaces and form a semicircular cross section on one edge. While you're at it, sand the ends smooth and draw a ½" circle on each, centered as close as possible in your 1"-diameter semicircle (Figure C).

Then drill each end with a ½" spade or brad-point bit (Figure D). The holes don't need to meet in the middle (as shown in the diagram). You'll hide your shiny new magnets in these holes, pressed in from both ends.

Next, draw cut lines for the overall depth of the holder (1") and the slots. The slots should be spaced approximately ¾" apart, with a depth of at least ¾" but no deeper than 3/16" (Figure E). The lines may be drawn on one or both sides of the holder, and the slot lines may be drawn on one side or all the way around (whatever works best for you in making the cuts). Then saw the holder to its 1" front-to-back dimension (Figure F).

3. Cut the slots for the knives.

Now, preferably with a band saw (Figure G), or alternately with a sharp hacksaw (Figure H), cut the knife slots 1/16" wide working from the rounded edge toward the back surface. Keep the slots straight and parallel and no more than 3/16" deep. Check the slot widths to be sure each just clears a knife blade.

Clamp the holder in a vise and gently file chamfers on each slot to help guide the knives in (Figure I).

4. Press in the magnets.

And now for the magnets! Important: They must be inserted in a certain order into the ½" holes, as shown in Figure J. Press them in with a short piece of undersize wood dowel (a 7/16" diameter works fine).

My magnets required a light press into the holes. If yours are more difficult to push in, you can gently tap them in with a hammer or redrill with the same bit. (If you bang them in, you'll break the holder because the slots

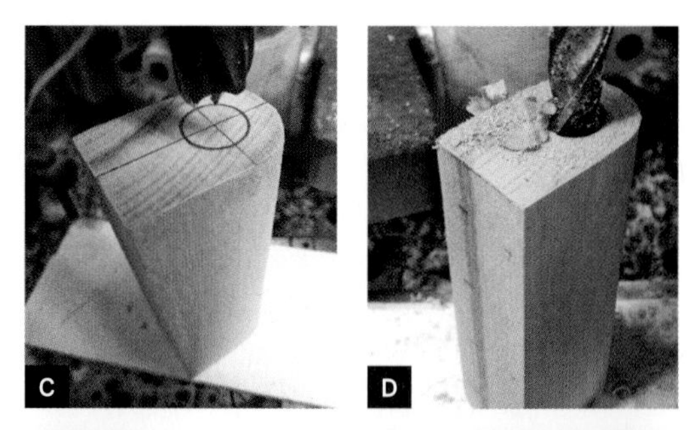

C D

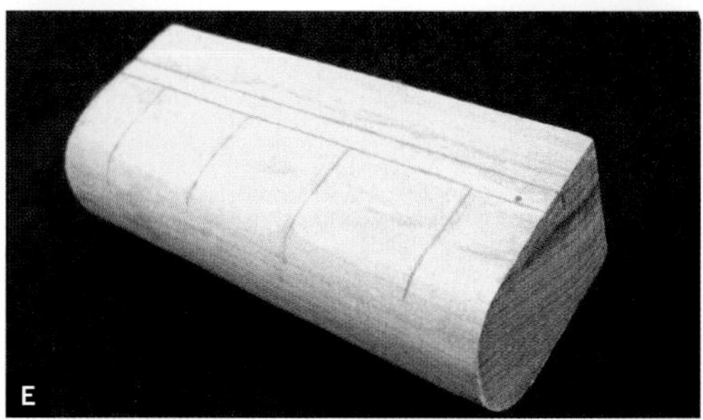

E

F G

H

I

weaken it somewhat.) If they're not a press fit at all and simply slide in, apply a little super glue around the rim before pushing the magnets quickly into place.

5. Make a backing board.

The backing board is made from ¼" plywood — the width should be the same as your holder and the length should be at least ¼" longer than your knives.

To determine the mounting location of the holder, place one knife in the holder with the top of the knife handle just touching the bottom of the holder. Adjust the holder so that the bottom of its handle rests about ¼" above the bottom edge of the backing board. Remove the knife and mark the location of the holder, then repeat this process on the back side (Figures K and L).

To mount the holder, start by marking the backside of the backing board along the holder's centerline, approximately 1" from each end of the holder (for a 3" holder, this can be adjusted to ½" from each end — the main idea is to stay clear of where the magnets and knife slots will be). Drill the holes using a #6 countersink.

Alternately, drill through-holes to clear the body of the screws, then countersink just enough to ensure the tops of the screw heads are flush (or barely sub-flush) with the back of the backing board.

Next, set the holder in position on the front of the backing board, and hold it in place securely using painter's tape or a couple dabs of hot glue. From the back, use the holes you just drilled as a guide, and drill through into the holder (make sure the holes are shallow and small enough for the #6×½" wood screws to grab well and still countersink). Screw the pieces together.

Sand the assembly and finish it with wax or stain. The holder is done.

To mount the assembly to the inside of a cabinet door, mark the front of the backing board an inch or two down from the center of each end, and pre-drill using a #8 countersink. Figure out where you want the assembly

⚠ CAUTION: These magnets are very powerful and potentially dangerous. Keep them away from children. They can break with rough handling. Please read the instructions that come with the magnets.

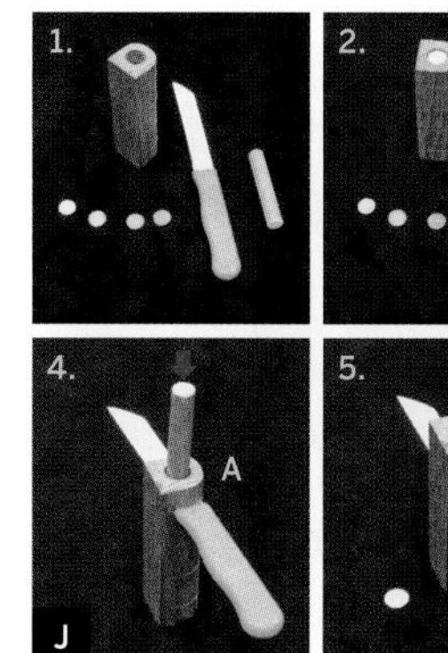
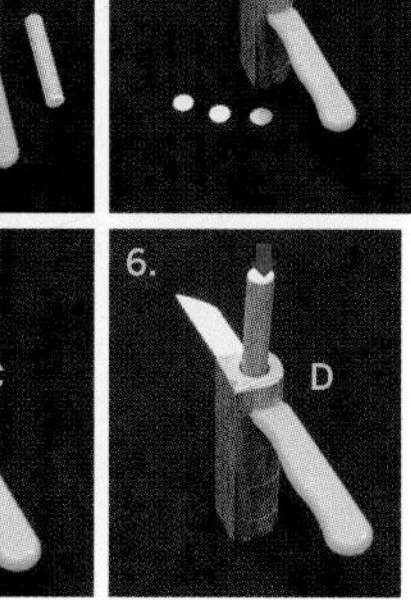

1. Ready to begin.
2. Put a magnet into the top hole.
3. Put a knife in slot B. Push the first magnet to the knife.
4. Put the knife in slot A. Push another magnet to the knife.
5. Flip holder and repeat for slot C.
6. Repeat for slot D.

mounted, mark and drill the mounting location, and attach the assembly using #8×¾" wood screws. You're done!

Larry Cotton is a semi-retired power-tool designer and part-time community college math instructor. He loves music and musical instruments, computers, birds, electronics, furniture design, and his wife, not necessarily in that order.

Simple Art Easel

Make an easy easel for your artwork.

By Liz Llewellyn

THIS EASEL IS A PERFECT WOODWORKING project for a beginner. The joints are simply glued, clamped, and screwed together, and the only power tool you need is a drill. The shelf is attached with pegs, using a technique that requires little precision and is impressively clean-looking.

1. Cut the pieces.

Following the diagram in Figure A, use your miter box and handsaw to cut the rear leg: one 61" length of 2×3 lumber, with perpendicular (90°) cuts.

For the front legs, cut two 58" lengths of 2×3, with angled cuts at each end; these cuts should be parallel to each other.

Now cut 2 lengths of 2×3 with angled cuts pointing toward each other (Figure B): the top brace 8¼" along its long edge, and the cross brace 31" along its long edge.

Make all the angled cuts at the same angle. The smallest angle on my miter box is 22.5°. This produces a very functional but somewhat wide-legged easel. If you have an adjustable miter box or an electric miter saw, 11° to 15° will give you a more compact easel that's easier to move around and set up.

If you use a smaller angle for the mitered cuts, the 2 braces will be a bit shorter. Lay the pieces out and eyeball the lengths. The top brace should be just large enough to mount the front legs as shown in Figure C, and the cross brace should be about halfway up the legs. Finally, cut two 3½" lengths of dowel.

Sand all the pieces well. The cut edges should be reasonably smooth, and everything should be free of snags and splinters.

Photography by Liz Llewellyn

MATERIALS

Dimensional lumber, softwood, 2×3, 8' lengths (3)
Nominal 2×3 lumber actually measures 1½"×2½". Inspect each piece for straightness and flaws; you want 3 nice clean lengths of 2×3.

Dimensional lumber, softwood, 1×6, 29½" long
A nominal 1×6 board is ¾"×5½", but any width between 4"–7" would work.

Strap hinge, about 3¼" wide × 4" long

Wood screws, flathead, #10×1¼" (6) or whatever size fits the countersunk holes in your hinge

Wood screws, #8×2½" (10)

Wood dowel, ⅜" diameter, 6" or longer

Small screw eyes (2) aka eye screws or eye bolts

Lightweight chain, 3'

Wood glue or polyurethane glue

Sandpaper

TOOLS

Handsaw ideally a straight backsaw

Miter box Mine has cam pins for clamping the workpiece in the box.

Drill with drill bits: ⅜" spade, 5/64" twist The 5/64" bit is for the wood screw pilot holes; you may want other sizes for screw eye pilot holes and a starter hole for the ⅜" bit.

Tape measure

Circular saw (optional) to cut the 1×6. You could also have it cut at the lumberyard.

Bench vise

Clamps (2) that open to at least 4"

Pliers (2 pair)

Electric miter saw (optional)

Framing square (optional) to mark cuts

2. Join the legs and braces.

Position the 2 front legs on top of the top brace as shown in Figure C. Dampen the mating surfaces if you're using polyurethane glue. Apply a thin coat of glue, and clamp them together until the glue is dry.

Once the glue has set, drill pilot holes and screw the pieces together with wood screws. Don't skip the pilot holes, or you'll end up splitting the wood. If you still end up with a split, glue the crack and clamp it shut while the glue dries.

Position the cross brace as in Figure D. To get it nice and level, carefully measure the distance from the bottom of the legs to the bottom of the brace. Glue and clamp, as you did at the top, and reinforce with screws.

Position the hinge on the top brace and rear leg so that the easel can fold completely shut.

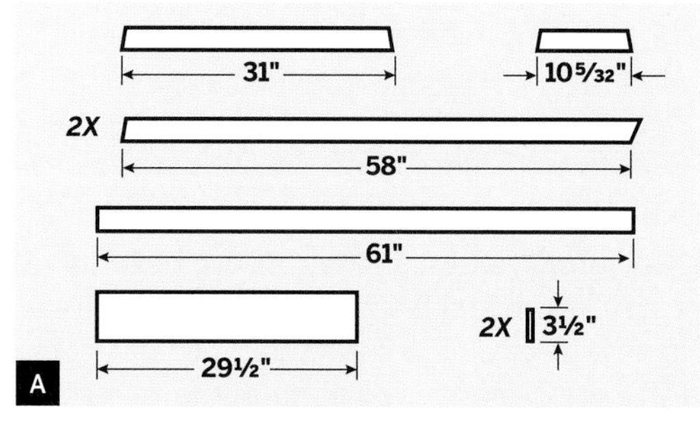

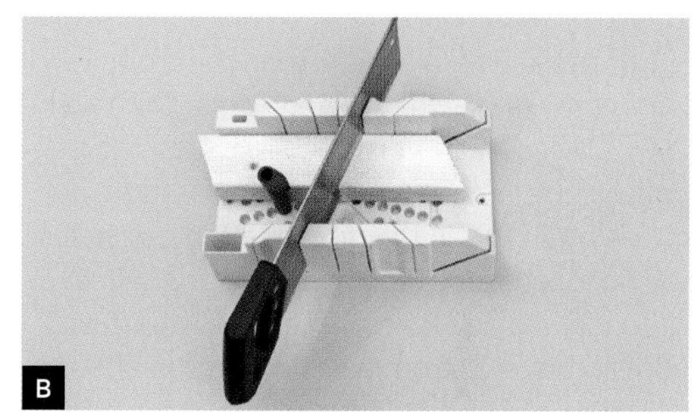

Drill pilot holes and attach the hinge with wood screws (Figure E, previous page).

3. Attach the shelf with pegs.

Drill ⅜" holes through the front legs for the shelf support pegs, about 1" above the cross brace (Figure F). Insert the pegs all the way into the holes so that 1½" sticks out the front (Figure G). They should fit tightly, but you can glue them if needed.

Cut the shelf a bit wider than the front legs, when level with the peg holes. Mine was 29½". Measure the distance between the peg holes, subtract this from the shelf length, and divide by 2. This is the distance the peg holes will be inset from the edges of your shelf. In my case, the peg holes were 25½" apart, so I drilled holes 2" from each end of the shelf.

Again using the ⅜" bit, drill the 2 peg holes into the shelf, 1½"–2" deep (Figure H), then check the fit by slipping the first hole onto a peg and seeing whether the second hole needs to be widened into a slot. If it does, drill a pilot hole next to your second hole (Figure I), then use the ⅜" bit to drill a third hole and drill out the excess wood between the 2 (Figure J). Don't worry if the slot is messy-looking; it'll be hidden by the front leg.

Slide the shelf onto the pegs.

4. Attach the chain.

Drill pilot holes for the 2 screw eyes: one in the center of the backside of the cross brace, the other on the inside of the rear leg, about even with the cross brace (Figure K). Screw the eyes in, then attach the chain to them by prying the last links open with pliers and then crimping them shut around the eyes.

Use It

If you're painting on paper, you'll need something rigid for support. I used a piece of plywood (Figure L), but cardboard or foamcore would work just as well.

Liz Llewellyn lives and works in small mill town near Boston. She designs embedded electronic devices with nifty mechanisms during the day and fabricates useful objects for herself, her friends, and her family at night.

F

G

H I

J

K

L

Anti Dog-Bite Siren

Hack an alarm siren to make a DIY dog repeller.

THIS TIME, INSTEAD OF BUILDING something new, I'm going to modify an existing product. I want you to see how quick and easy a mod can be, if you have just a little electronics knowledge.

The product I'm interested in is an alarm siren. This suggests something you'd find on an emergency vehicle, but in the world of burglar alarms, a siren is a little gadget that usually makes a high-pitched squealing sound at 2kHz or above (2,000 cycles per second).

My goal was to adjust the siren to an even higher frequency, to discourage unfenced, aggressive dogs that chase my mountain bike in the rural area where I live. I must emphasize that I'm an animal lover, with no desire to cause pain or injury. I just wanted to say, "Back off, pooch!"

I had already tried an ultrasonic dog repeller that I bought from a catalog, but it didn't work very well. On the few occasions when I tested it, a dog sometimes paused and gave me a funny look — but then continued to chase my bicycle anyway.

Because a dog repeller operates above human hearing range, I had no way of knowing how loud mine really was. I suspected it wasn't loud enough, because when I peeked inside, I found that the transducer (which converts electricity to audio) was quite small.

Also, I didn't think the repeller was sufficiently annoying. Even dog hearing is less sensitive to extremely high frequencies, and maybe an elderly dog wouldn't be able to hear ultrasound at all.

I decided I wanted a frequency that was audible to me, while being louder to dogs.

MATERIALS

Alarm siren, 12V DC, 2.0kHz–4.0kHz PUI Audio part #XL-5530L-LW300-S-R, mouser.com, $21
Resistor, ¼W, 68kΩ
Switch, SPST or SPDT toggle, subminiature
Battery, 9V
Battery connector, 9V
Velcro, adhesive-backed
Masking tape
Electrical tape, or heat-shrink tubing and heat gun
Epoxy glue, quick-drying

TOOLS

Hacksaw, Phillips head screwdriver, hobby knife, wire cutters, needlenose pliers, alligator clips, soldering iron and solder, earplugs or other hearing protection, magnifying glass (optional)

⚠ CAUTION: Wear hearing protection. Out of the box, when powered by 12V DC, the PUI unit is rated to create noise fluctuating between 2kHz and 4kHz, with a sound pressure of 123dB at a distance of 10cm. This is frighteningly loud; I'm concerned it could damage your hearing if you expose yourself to it for a long time, at close range, in a room where walls and ceiling reflect the noise. It should be safe once you're outside, but please wear hearing protection while working on this project. Foam earplugs are sold at hardware stores, and they're good safety equipment to have in any workshop.

Something between 5kHz and 10kHz might do it.

I could have built my own device from scratch using individual components, but when I looked at transducers in catalogs, they were mostly rated at less than 100 decibels (dB). A pre-built siren was much louder, so why not work with that?

If you've ever hesitated to pry open a product for fear of damaging it, here's how it can be done.

A

Fig. A: The alarm siren, straight out of the box.

Fig. B: Secure the metal mounting bracket in a vise and apply a hacksaw, using tape to guide your cut. Don't let the blade penetrate deeper than this.

Fig. C: Snip the audio output wires.

Fig. D: Unscrew the mounting bracket and carefully cut away sealant around the power input wires. It doesn't matter if the wires are damaged. You can replace them later.

Fig. E: Remove the screw to release the circuit board.

B

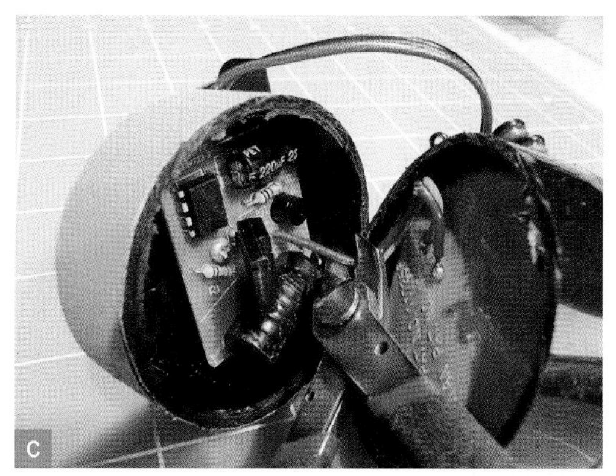

C

→ START

Opening It Up

The siren I chose is made by PUI Audio Inc. You can find the data sheet at puiaudio.com, and you can buy it from mouser.com or other suppliers for $20 to $25. That's cheaper than a dog repeller, and the mods shouldn't add significantly to the price.

The PUI siren comes in a sealed plastic case (Figure A). To get inside it, I selected a fine-toothed hacksaw. The big question was where to cut. In this situation I always ask myself: if I designed this thing, where would I put the components? In this siren I would probably attach a little circuit board either directly beneath the transducer or in the bottom of the case. With this in mind, I started cutting midway between these extremes. I used bright green masking tape to guide my cut, and barely allowed the blade to penetrate (Figure B).

I found the electronics in the bottom of the case, with red and yellow audio output wires leading to the transducer. I clipped these wires (Figure C), then unscrewed the metal mounting bracket from the bottom of the case and used a utility knife (Figure D) to cut away sealant where the power cord entered. Be very careful if you use a knife like this. Always cut away from you!

D E

After removing just one internal screw (Figure E), I could pull out the circuit board. Would it still work? I reconnected it with patch cords and tested it with a 9V battery (Figure F, following page). Everything was still good. Now what?

Photography by Charles Platt

Testing a Mystery Circuit

The siren's circuit contained 6 simple components: 3 resistors, 2 diodes, and a 220μF capacitor. In addition there were 3 mystery parts: one that looked like a coil (with no markings), an 8-pin chip, and a 3-pin thingy that was maybe an output transistor.

I guessed that the 8-pin chip was a microcontroller, because there weren't enough other components to create the two-tone squealing sound. I couldn't find its part number when I searched online.

At this point I could have drawn the circuit by following the copper traces underneath the board, but so long as the 8-pin chip was a mystery, a circuit diagram probably wouldn't help me much. So, I went back to first principles. I knew that circuits that oscillate are often controlled by resistors and/or capacitors. (How would you know this? Maybe by reading a copy of my book, *Make: Electronics*!)

I selected a 100μF capacitor and touched it across the 220μF capacitor while the siren was running. I was disappointed to find that this made no difference at all. (Incidentally, the presence of a coil on the circuit board suggests that this device may crank up the internal voltage somewhat. If you work with it while the battery is connected, you may want to use insulated pliers.)

The capacitor experiment was a failure, but what about the resistors? Their colored bands told me they were rated at 1K, 470 ohms, and 200K. I guessed that 200K was the most likely to control an audible frequency.

I added various standard-value resistors to the solder pads under the 200K resistor, using a 30-watt soldering iron (because the kind of 15-watt iron that I normally use wasn't powerful enough to melt the pads). When each resistor was attached, I tested the alarm and found that the frequency changed. I got precisely what I wanted when a 100K resistor seemed to double the frequency (Figure G).

This was very encouraging, but for my final mod, I needed something neater. I could remove the original resistor and my add-on resistor, and substitute a single resistor in their place — but what should its value be?

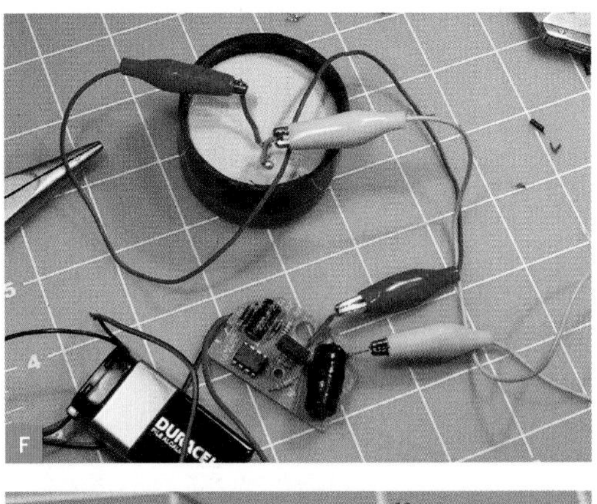

F

G

Making the Mod

If you have 2 resistors, with resistances R_1 and R_2, wired in parallel, their overall or total resistance, R_T, is given by this formula:

$$\frac{1}{R_1} + \frac{1}{R_2} = \frac{1}{R_T}$$

In the siren, R_1 was 200K and I had added 100K as R_2. So,

$$\frac{1}{200K} + \frac{1}{100K} = \frac{3}{200K} = \text{approx } \frac{1}{67}K$$

In other words, I could substitute a single 68K resistor for the pair that I had tested.

But wait a minute. Would the lower resistance burn out the siren? Well, an alarm siren is designed to run for very long periods, whereas I was planning to use the modified version only in short bursts, and at 9V instead of 12V. Under these conditions, I guessed that a resistor of ⅓ the original value should be OK. The 9V battery would probably fail before the siren did.

OPPOSITE

Fig. F: Reconnect the audio with patch cords, and use a 9V battery to test that the unit still works.

Fig. G: Add a 100K resistor temporarily to increase the audio frequency.

THIS PAGE

Fig. H: Unsolder the 200K original resistor and thread in a 68K replacement.

Fig. I: With a 9V battery connector, a submini on-off switch, and wire joints insulated with heat-shrink tubing, the unit is ready to go back in its case.

Fig. J: The modified siren, with on-off switch and a battery stuck to the bottom with velcro.

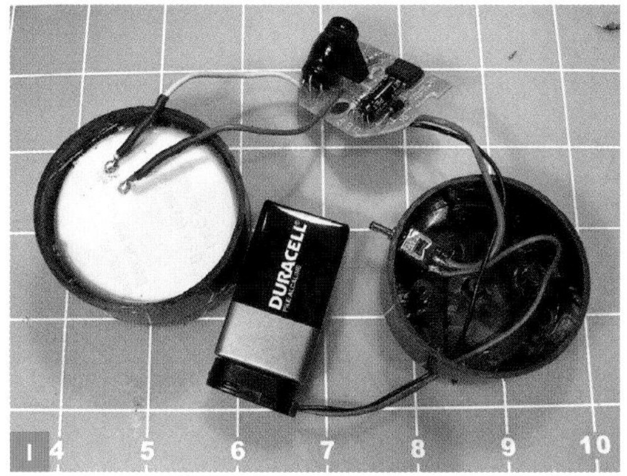

I removed my 100K resistor, then melted the solder pads and pulled out the original 200K resistor with needlenose pliers. I substituted my 68K resistor (Figure H), retested the siren, and, yes, it still ran at the higher frequency. In fact, after running it, I heard dogs barking a quarter-mile away — even with all my windows closed!

I glued a subminiature on-off switch through a hole in the case, ran in the wires from a 9V battery connector, and routed one wire directly to the circuit board while the other went to the board through the switch (Figure I). I used adhesive-backed velcro to stick a 9V battery on the outside of the case, then secured the sawn halves using red vinyl tape (Figure J).

Field Testing the Dog Siren

Finally, I went out on my mountain bike until I found a potentially troublesome dog. I used the modified alarm siren, and the dog looked — well, alarmed! It took a step back. I gave it another burst, and was happy to find that I could continue on my way without being chased and harassed.

I'm wondering, now, if I should mount a second siren on my car, to discourage deer and elk from leaping in front of me at night on Interstate 40. This is a life-threatening problem in my area, and although a screaming noisemaker is an unusual auto accessory, I can't think of any law against it — so long as it doesn't sound like a real emergency siren, of course.

As for the moral of this story, I hope that it's now obvious. Never be afraid to make mods.

Make: Electronics book at the Maker Shed: makezine.com/go/makeelectronics

Charles Platt is the author of *Make: Electronics*, an introductory guide for all ages. A contributing editor of MAKE, he designs and builds medical equipment prototypes in Arizona.

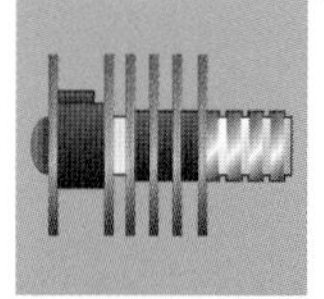

MAKE MONEY
By Tom Parker

Sometimes it costs more to buy it than to make it from the money itself.

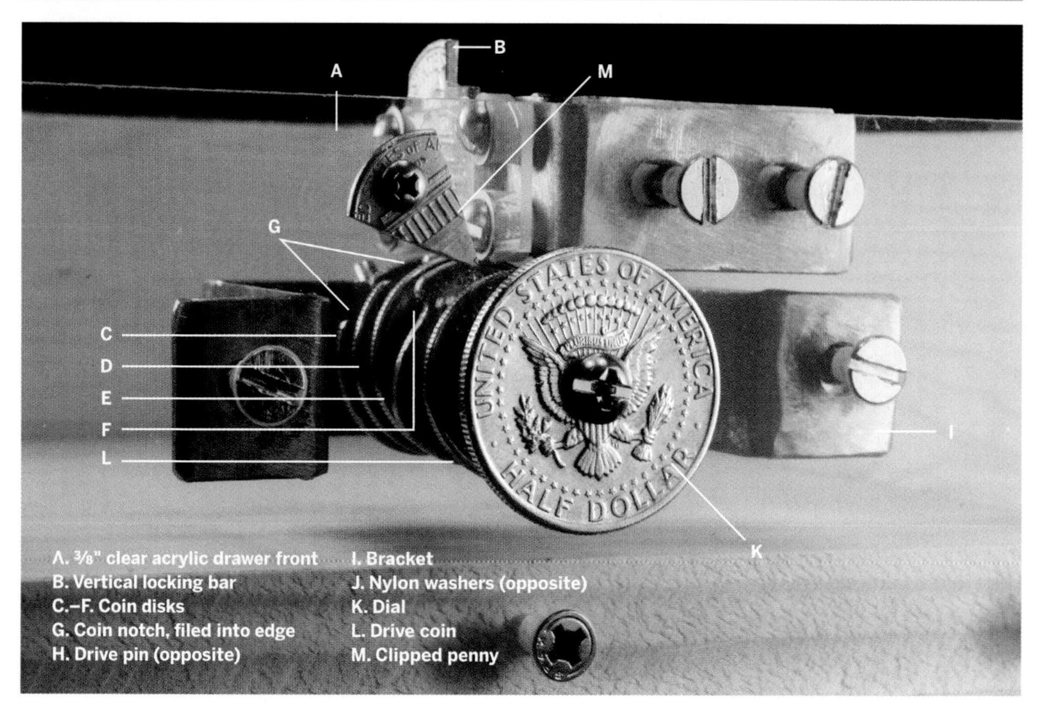

Combination Lock

SIMPLE COMBINATION LOCKS ARE based on a shaft with rotating disks. I made this one out of 5 quarters, 2 half-dollars, and a penny (plus an odd assortment of scrap metal, nuts, and bolts). It looks complicated but it's actually very simple.

For illustration purposes, I made the drawer front out of ⅜"-thick clear acrylic (A) so you can see both inside and outside the drawer.

The drawer lock works like this: when the drawer is closed, a vertical locking bar (B) made from one of the 50-cent pieces is moved upward by a cam filed into coin (F), engaging the roof above the drawer. This prevents the drawer from sliding open.

Four coin disks — (C), (D), (E), and (F) — with holes drilled in their centers are aligned on a coin shaft below the locking bar. These coins collectively hold the locking bar in place and keep the drawer locked.

But each coin also has a notch filed into its edge (G). It doesn't matter where on the coin the notch is filed, but its location becomes "12 o'clock" for the purposes of installing the drive pins (H) later. Aligning all 4 of these notches under the locking bar lets the bar drop, unhooking itself from the roof above the drawer. This lets the drawer slide open! (The exact notch dimensions, hole sizes, and pin widths depend on what sort of hardware you have on hand.)

So how do you turn the 4 notched coins? The coin shaft is mounted with one bolt to a bracket (I) on the inside of the drawer (the locking bar is bolted to its own separate bracket). This coin shaft protrudes from the bracket toward the inside front of the drawer. The coins are separated with nylon washers (J) so that each can rotate independently on the coin shaft. In addition to a notch at

Photography by Tom Parker

$15–$20

Combination drawer lock bought online.

$2.26

Combination drawer lock made from coins.

C.–F. Coin disks
G. Coin notch, filed into edge
H. Drive pins
J. Nylon washers
L. Drive coin

12 o'clock, each disk has a short "drive pin," made from metal rod or tiny bolts. Coin (C) has a pin at 11 o'clock, (D) at 7 o'clock, (E) at 4 o'clock, and (F) at 2 o'clock.

The dial shaft is mounted to the front of the drawer and aligns with the coin shaft. The 50-cent dial (K) outside the drawer turns the drive coin (L) inside the drawer. The drive pin on coin (L) then engages the drive pin on coin (F), which engages the drive pin on (E), and so forth, so that all the notches can be rotated and aligned — combination lock style — allowing the locking bar to drop out of engagement with the roof above the drawer.

Note that one side of the notch in coin (F) is filed out to form a cam. When the drawer is reclosed, reversing the direction of the dial causes this cam to lift the locking bar back into the locked position in the roof above the drawer.

A final detail, the clipped penny (M), is added to the front of the drawer as a registra-

tion mark for entering the combination.

Once you've built and tested your lock, you can adjust its orientation to the dial for an interesting combination, like "AL – M – ST – Eagle Head." Who says the dial on a combination lock has to use numbers?

Tom Parker (parker@rulesofthumb.org) lives in Ithaca, N.Y., and works for Cornell University. When he's not tinkering with junk, he flies a 1956 Cessna 180 bush plane.

TOYS, TRICKS, AND TEASERS
By Donald Simanek, recreational physicist

Kinetic Illusion Toys

Deceive the eye with impossible falling motion.

VISUAL ILLUSIONS COME IN MANY FORMS. The most common are flat pictures, cleverly made so they seem to show something that's impossible in reality. A smaller category is the kinetic illusion, an illusion caused by motion of real objects.

Jacob's Ladder

Let's look at two kinetic illusion toys, one well-known, one not so well-known. Both give illusions of something continuously falling that doesn't really fall.

The folk toy called "Jacob's ladder" is also known as "magic tablets," "Chinese blocks," or "klick-klack" toys. We don't know exactly when or where it originated. It's a simple toy consisting of blocks of wood strung together with ribbons. When suspended from one end, the top block is turned over, and a block appears to slowly fall to the bottom, flip-flopping with a nice clattering sound (Figures A and B).

The first documented description of this toy is from *Scientific American* in 1889. Not many toys have generated the historical mythology this one has. Some say it has its roots in China, others that it was found in King Tut's tomb. But the contents of that tomb were carefully inventoried when it was discovered, and no mention was made of finding anything like this toy.

The name "Jacob's ladder" refers to the biblical passage in Genesis 28:12 in which Jacob dreams of angels ascending and descending a ladder or stairway that leads from Earth to heaven. It's been said that Jacob's ladder was one of the few toys some Puritan households allowed their children to play with on Sundays because of its biblical reference. But the true origin of the toy remains a mystery.

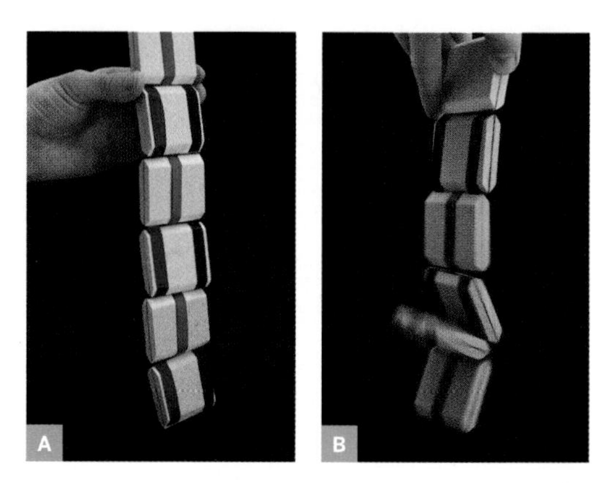

Fig. A: Jacob's ladder toy.
Fig. B: Jacob's ladder in action.

Over the years, minor variations of Jacob's ladder have been frequently patented in the United States. Commercial versions consist of colored blocks strung together with colored ribbons. Those with blocks of mixed colors, in my opinion, compromise the illusion of the falling block.

In the 1940s, when I was a child, these toys were sold with a round indentation in one block that had a penny nested in it, so you could perform a little trick to make the penny seem to vanish. Its block just flipped when the block stack was unfolded, so the side with the penny was on the bottom side, secured by the center ribbon. Those were the days when, for a kid, a penny was worth having. (A folded dollar bill works even better.)

The ribbons used to construct these toys can be narrow or wide, but I suggest good-quality, strong, non-stretching cloth ribbons

Photography by Sam Murphy

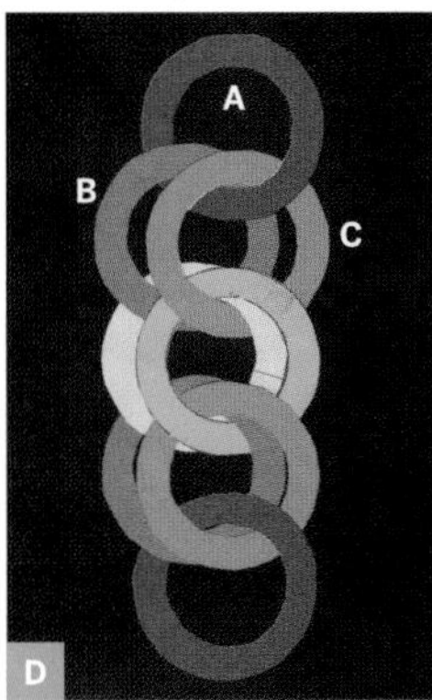

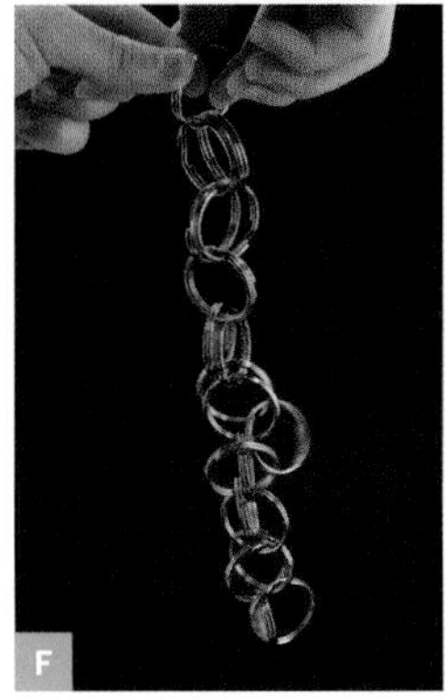

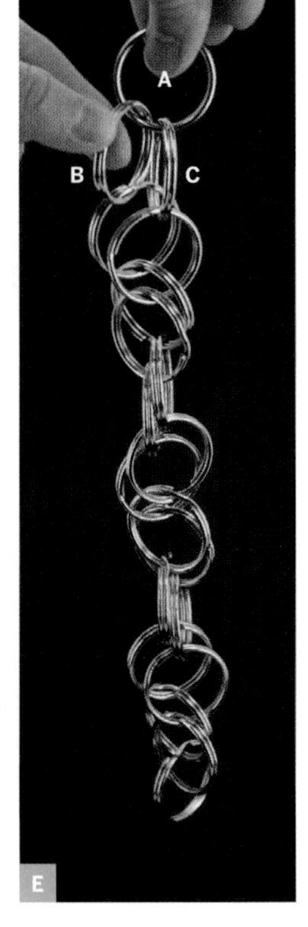

Fig. C: Ribbons are secured between the halves of each block.

Fig. D: Construction diagram for the tumbling rings. The colors show how the rings must be linked.

Fig. E: The two interlinked chains of rings. Ring B is about to become the top ring.

Fig. F: The tumbling rings in action, with the "falling" ring halfway down.

for durability. The blocks in some commercial sets have rounded edges, which reduces ribbon wear and gives the smoothest operation. If squared-edge blocks are used, the ribbons can be secured with small tacks or glue (only at the ends of each block).

Figure C shows one of the best designs, with each block made of two wooden slabs glued together, and the ribbon ends secured between them.

As you rotate the top block, you keep it at a constant height. The visual impression is that the block tumbles all the way to the bottom, especially if the blocks are the same color and only one color of ribbon is used. In action, each block just rotates 180° and doesn't fall at all.

The Tumbling Rings

The first description I saw of this illusion was in a 19th-century book of magic tricks for entertaining. It later appeared in *Scientific American*, in Martin Gardner's "Mathematical Games" column in the 1950s. Formerly called "Afghan rings," these assembled chains may be purchased at magic shops.

For the tumbling rings illusion, an even number of rings is interlinked as shown in Figures D and E. I used split key rings, which can be found in several sizes in craft or hardware shops.

If you hold the top ring (ring A), you can test the two rings just below by trying to lift them. Ring B can be lifted so it will cause the half-chain below to lift with it. Ring C lifts the whole chain — ignore it.

Hang onto ring B and let go of ring A. You may be rewarded by a smooth "falling" action, which appears to onlookers like the top ring fell all the way to the bottom.

Actually nothing falls. The other side of the chain simply untwists, giving an illusion of falling. If this doesn't happen, you've grasped ring B on the wrong side (try grabbing the other side), and thus dropped A in the wrong direction (try the opposite direction). If the tumbling illusion "hangs up" on the way down, you've interlinked some rings incorrectly.

Sometimes it helps the illusion to release the top ring with a bit of rotational flourish, giving it initial angular momentum. Once you get the hang of it, you can keep a tumbling action going by transferring your hold from the top ring to the next after each fall, timing your action to the period of falling.

Photograph by Donald Simanek (D)

At any step, there are four options for grasping a ring below the top ring; three of these are wrong. Suppose ring B is on your left side. Grasp it at the point nearest you, and drop ring A forward (away from you). Then the next time, ring B will be on your right side — grasp it at the point farthest from you and drop ring A backward (toward you). Keep alternating this action.

Words can't convey the hypnotic fascination of watching the tumbling rings illusion in action.

Figure E on the previous page shows the full chain as you'd see it in your own hands, with ring A being held at the top and ring B on the left about to be lifted. Grab ring B on the near side and lift it, as you release ring A away from you (again, ring C is the wrong one).

Figure F shows the "falling" ring in action, halfway down the chain.

Gardner's description of this illusion inspired me to make one for myself, and I hope you do, too — a verbal description can't convey the hypnotic fascination of watching it in action. (Read an interview with Gardner in MAKE Volume 12, page 80.)

Shortly after his column appeared, I went to a key-making shop to get the rings — 20 of them. The proprietor said, "You must have a lot of keys." So I explained what I intended to do with them. He hadn't heard of anything so daft, but he must have sensed my enthusiasm for the project, as he ended up giving me the rings at a substantial discount. Those were different times.

The actual assembled chain does not lie flat, which makes it difficult to illustrate with a drawing. In fact, some published drawings show improper linkage that could never work. Use Figures D and E as your guide. Notice that pairs of rings naturally lie side by side when suspended, as shown in Figures E and F.

When you assemble the length you want, or are about to run out of rings, just link a final ring through the bottom two, and the chain is finished. The chain will have topological symmetry top to bottom. It will look and operate the same whichever end you suspend it from.

A chain of 2" key rings is easily carried in a pocket. I've made larger sets using plastic macramé rings from craft stores, the kind that can be opened at a locking socket, linked, and relocked.

I've also made a set using craft-store steel rings. Half the rings must be sawed open at the weld point with a hacksaw. Assemble the ring chain, weld the breaks closed, and file off any irregularities from the welds.

In terms of a good performance, 5" rings are about the size limit. Larger sets fall more slowly, and the illusion isn't as effective. The number of rings used is determined by the size of the rings (you don't want the chain dragging on the floor when you demonstrate it!). For the 2" key rings, an even number somewhere between 20 and 30 works best.

In action, it appears that the top ring is released and cascades all the way to the bottom. But in fact, you're just exchanging the top ring with the next one down the chain, and half the chain twists 180°, starting at the top. You can easily confirm this by taping a marker on the top ring, if you have any doubts.

A nice touch when showing this to an audience is to finish by inverting the chain and demonstrating that it works just as well in reverse. Tell them you're restoring the chain to its original condition, putting the rings "back where they belong," and thereby restoring the order and harmony of the universe.

I've never known anyone who could assemble one of these after simply observing it in action. Even with the instructions here, most people struggle to link the rings correctly.

Visit makezine.com/26/ttt for a vanishing trick with Jacob's ladder.

See a video showing how to make the tumbling rings at makezine.com/go/tumble.

Donald Simanek is an emeritus professor of physics at Lock Haven University of Pennsylvania. He writes about science, pseudoscience, and humor at www.lhup.edu/~dsimanek.

 By Lee D. Zlotoff

Charge That Cell!

The Scenario: While on an early winter camping trip you get caught in a snowstorm, which you handily survive (*see Volume 22, "MakeShift: Snowbound"*). But when you finally hike out and return to your car, you discover that from the drop in temperature and your extended stay, your car battery is completely dead. What's worse, even though you had a solid signal when you first arrived in the area, your cellphone battery is now also dead from the cold. But you recall seeing a house just off the trail when you first hiked in, so you head over there to check it out.

At the house you try ringing the doorbell, but it makes no sound and you get no response to your knocking. You find a key under the mat, however, and you let yourself in. You notice that the room is still warm from the gas-powered wall heater. Clearly, whoever was here must've just left after realizing the electricity was out — no doubt another casualty of the recent storm. So you rummage around and find a toolbox with a good assortment of tools including one full propane torch, some lamp wire, and a kerosene lamp — but no batteries of any kind, anywhere. You know that if you can just figure out a way to recharge your cellphone, you can call for help to get your car going again. But can you?

What You've Got: In addition to everything mentioned above, you have all your camping stuff: tent, sleeping bag, camp stove, matches, and your Swiss Army knife or Leatherman tool. (Alas, your flashlight batteries are shot, too.) But will all that be enough to fire up that magic communicator in your pocket? You tell us. Good luck!

Send a detailed description of your MakeShift solution with sketches and/or photos to makeshift@makezine.com by Aug. 26, 2011. If duplicate solutions are submitted, the winner will be determined by the quality of the explanation and presentation. The most plausible and most creative solutions will each win a MAKE T-shirt and a *MAKE Pocket Ref*. Think positive and include your shirt size and contact information with your solution. Good luck! For readers' solutions to previous MakeShift challenges, visit makezine.com/makeshift.

Lee David Zlotoff is a writer/producer/director among whose numerous credits is creator of *MacGyver*. He is also president of Custom Image Concepts (customimageconcepts.com).

Photograph by Jen Siska

$$(\text{prototyping} + \text{engineering})^{\text{imagination}} = \text{better future}$$

$$\left(\frac{\text{dreaming}^2 + \text{experiment}}{\sqrt{\text{failure}} \times \text{renewal}}\right) = \text{invention}$$

$$\text{passion} + \text{education} + \text{hard work} > \text{complacency}$$

SPOOL RACER HOWTOONS.COM

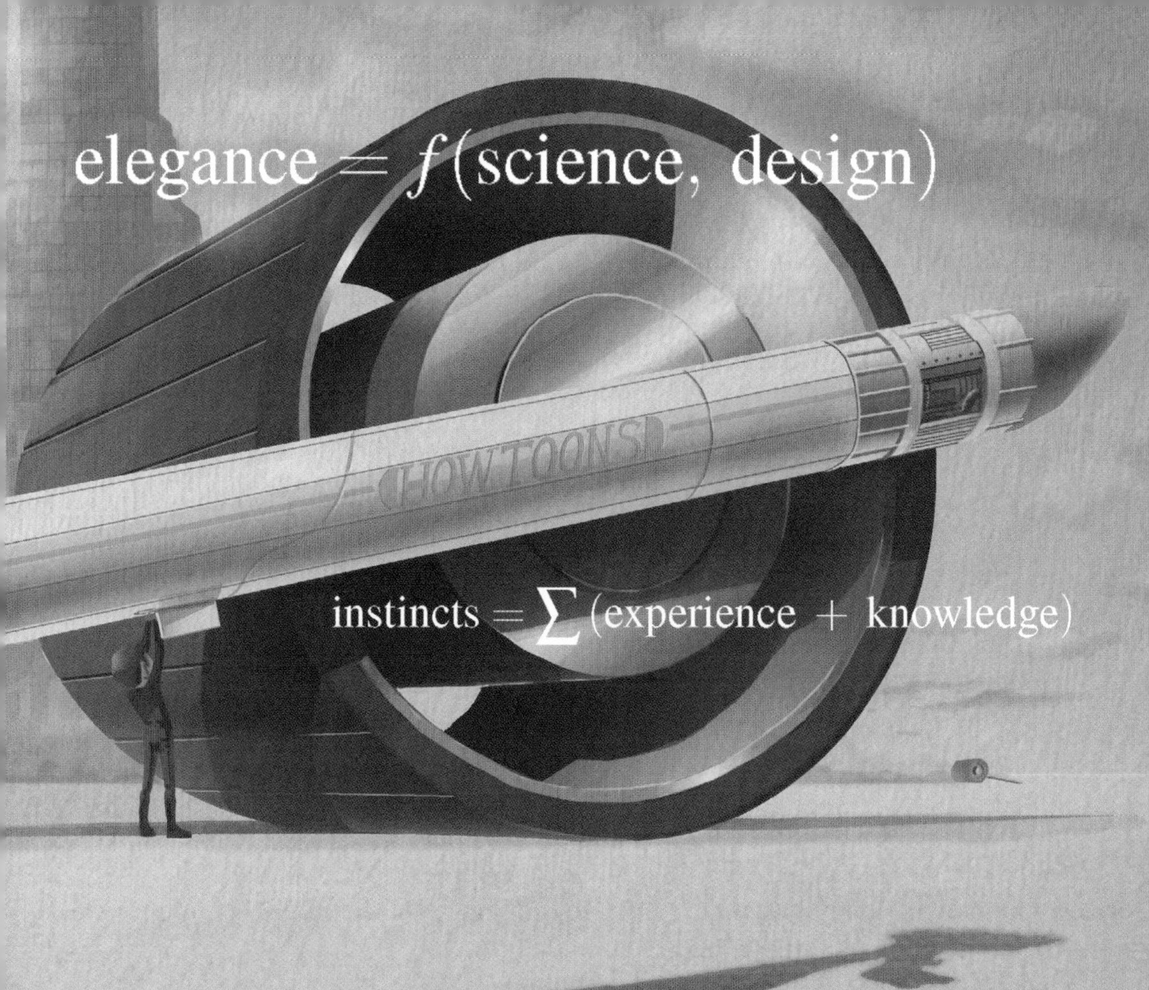

$\text{elegance} = f(\text{science, design})$

$\text{instincts} = \sum(\text{experience} + \text{knowledge})$

$\int \text{hard work } d \text{ opportunity} = \text{success}$

$\forall \text{ problems } \exists \text{ a solution } \in \text{brain}$

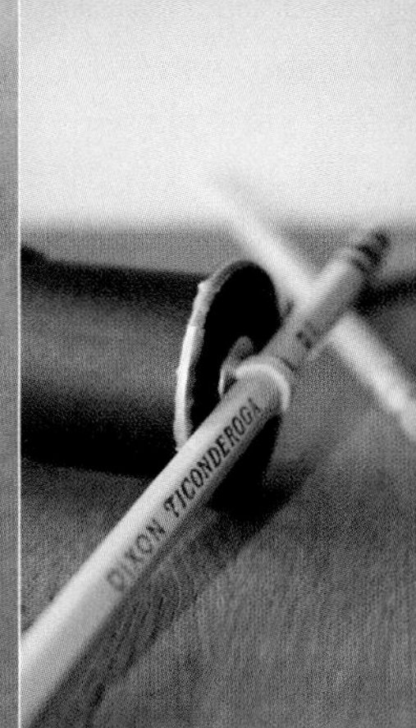

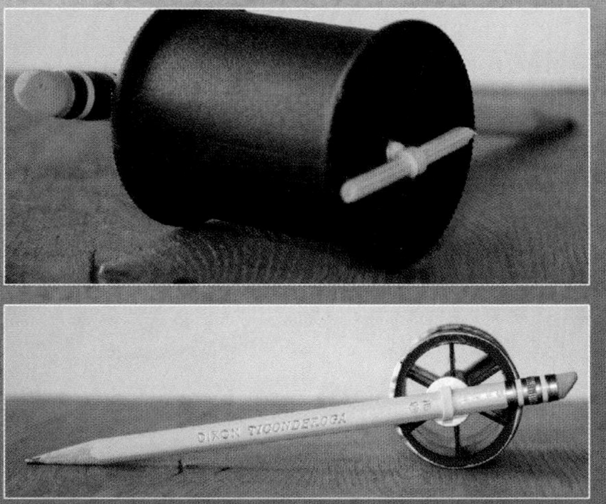

DRAGOTTA 2011

Hot stuff, from thermometers to a pocket hotspot; and rugged gear, from backpacks to drills to tweezers.

TOOLBOX

Drills and Drivers

Milwaukee 18V Compact Drill and Impact Driver Combo Kit

$225–$550 milwaukeepowertoolsonline.com

Believe it or not, there was a time before handheld electric drills. I grew up twisting my wrist and torquing my hand getting screws in and out of wood and metal. My father's fingers are actually bent from screwing so many bolts and screws by hand in his youth.

No longer are these indignities required, and a staple tool of any modern maker is the electric drill. I've owned my share, but I don't think I truly learned the beauty of an electric drill until I started using Milwaukee's 18V hand drill and its oft-ignored cousin, the 18V impact driver. I wouldn't do without either.

The impact driver is truly astounding for what it can do. I recently turned my attic into a playroom with it, and it gave me nothing but joy. If you don't know what an impact driver is, you should. It's kind of like a front-end loader — you probably think it's a luxury and too much tool for your job, but as soon as you have one, you search out jobs to do with it because it's just plain fun.

The Milwaukee 18V series feels like it's built as tough as a front-end loader, too. Which is exactly why I've managed a love affair with Milwaukee that was missing with other brands. It's just that special something of a tool that's better.

Yes, you'll pay more, but it's absolutely worth it. The better battery charge indicator is nice, but the real difference is in the quality of the chuck, the gearbox, and the overall build of these tools. Enjoy.

—Saul Griffith

BlinkRC Wi-Fi Receiver

$125 makershed.com/blinkrc

This tiny little circuit board is stuffed to the brim with components. It allows anyone to easily swap out a standard R/C receiver and communicate on a wi-fi network so you can control your R/C vehicle with your smartphone. Just plug it in (replace your stock receiver), download the free app, and you're off!

For the more adventurous, try creating an app that takes advantage of the open messaging protocol and the BlinkRC's three output channels and two analog input channels. Now you can control a variety of different servos and sensors from almost anywhere.

—Marc de Vinck

Spy Video Trakr

$130 spygear.net

Spy Gear offers all kinds of "spy technology" in toy form — all of which actually work. With Trakr, they've come up with a wonderfully robust toy robot. It's remotely operable, with the controller packing a speaker and screen so you can see and hear what the robot does, even in a different room. Built-in night vision, audio AV recording, robot-mounted speaker, and built-in SD slot add to the options.

You can program and upload apps from any computer, allowing you to create a complex series of actions that the robot performs autonomously. While this programmability intrigues, it's the robot's hackability that will grab your attention. The circuit board is visible beneath a clear canopy and has a ton of labeled empty pinouts. Everything is secured with regular Phillips screws, just begging you to crack it open.

—John Baichtal

PockeTweez

$25 pocketweez.com

I gave myself a splinter for this review. On purpose. It was that or focus only on the personal-grooming uses of these clever folding tweezers. And while the PockeTweez is definitely up to the job of plucking that hair discovered peeking out of your ear before the big date, it's really in splinter removal that the pointed-tip shape excels.

Although it's only been on my key ring for a week, I'm very optimistic about the design's pocket-hardiness. There are only two components: handle and tool, both cast stainless. The tweezers are machined to shape, and the handle is bead-blasted. The one locks into the other, in both folded and open positions, using a clever integral slot-detent with no additional parts to wear out or wander off. And it pulled out my test-splinter in a jiffy. Good thing, too, or I'd be typing one-handed.

—Sean Michael Ragan

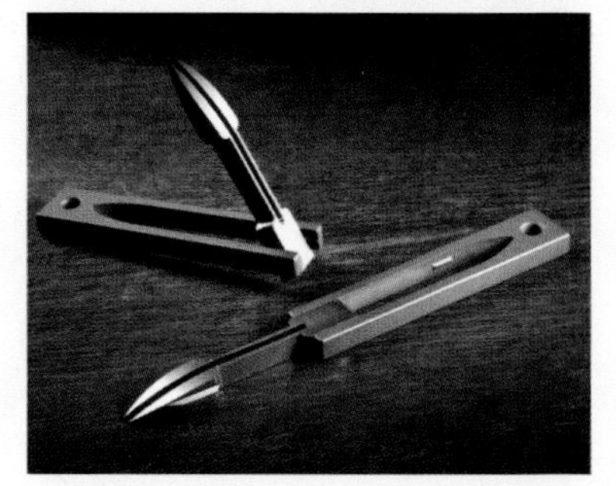

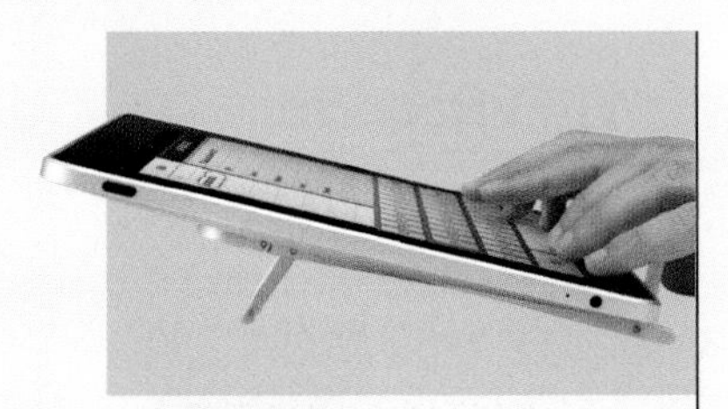

Compass iPad Stand

$40 twelvesouth.com

I use the heck out of my iPad and usually love its handheld form factor. There are times, however, when I need it propped up in front of me, hands-free.

This is usually when I'm using it as a reference guide in my workshop, cookbook in the kitchen, surrogate movie screen on an airplane, or mobile workstation in a hotel room paired with a Bluetooth keyboard. I've tried a few stands and even built my own, but my new favorite is the Compass stand from Twelve South.

This easel-style stand is very compact when folded up — a boon when traveling — yet it's a solid, no-compromise stand. It's made from powder-coated steel and has nonskid silicone on its feet and at all points where it touches your iPad.

When opened, it can be positioned in either of two modes: fully upright for reading and movie watching or nearly flat for an onscreen keyboard typing position.

My only complaint is that the sturdy little thing seems to draw attention when X-rayed by airport security guards. I've learned to take it out of my bag and put it in a tray. That's a small inconvenience, however, for an otherwise excellent accessory. *—John Edgar Park*

Eureka! By Roy Doty
Cookie Alarm

Rover Puck

$150 rover.com/home.htm

The Rover Puck mobile 4G hotspot is small enough to fit in a hip pocket. There's no contract required; once you've bought the device, you simply pay as you go for coverage (from $5 a day to $50 a month). Rover claims download speeds of 3Mbps–6Mbps and up to four hours of battery life; up to eight devices can connect at once. Coverage is limited to certain U.S. cities, so be sure to check Rover's online coverage map before you buy. *—JB*

One Ring to Rule Them All

MoMA Key Ring Organizer
$12 momastore.org

I'm often lukewarm on "modern" design. While I like to see clever products that put function first, the design still fails, in my opinion, if it has to be priced as a luxury item. So I love to browse the Museum of Modern Art catalog but rarely buy anything. The Key Ring Organizer, a staple of the MoMA gift shop for at least a decade, stands out as a shining exception: I bought four of them in 2001.

I kept one for myself and distributed the others to my family. We all still use them, and they all still function perfectly, after ten years of continuous pocket- or purse-wear per unit. Whenever we have to swap keys, it's literally two clicks to make the transfer from one ring to another. And at 12 bucks, the Key Ring Organizer is one of the cheapest items in the catalog. *—SMR*

STM Convoy

$100 stmbags.com.au

You know those flimsy nylon backpacks that never seem to last more than a single season? Time for something more rugged. The Convoy is made of thick, water-resistant canvas lined with nylon and is covered in pockets. Like any serious pack it has a built-in sleeve that's able to accommodate up to a 15" laptop. There's also a pull-out rain cover you can use if you get caught in a downpour.

—JB

KID REVIEW

Altium NanoBoard 3000 Series

$395 altium.com

The NanoBoard 3000 is a great electronics development tool. Installing and licensing the Altium software was probably the most challenging part of setting it up.

Altium makes creating schematics easy with its library components and user-friendly GUI. The OpenBus palette offers a range of "puzzle pieces," allowing you to connect the peripherals (touchscreen, speakers, etc.) without having to get down and dirty and write several lines of commands.

With Altium's software, I was not only able to create a schematic, but with a bit more work and time, I was able to actually test my project on a device rather than having the software debug it. Prototyping has totally changed for me! The NanoBoard can be used by a wide range of people, from hobbyists with a good grasp of computers and electronics to computer-electronics professionals.

—Max Rohde

Thermapen Meat Thermometer

$89 thermapen.com

While we were cursing over our oven, trying to figure out if a roast chicken was done, our engineering-geek guests were rolling their eyes.

A few days later, a Thermapen showed up in the mail, a generous gift given out of pity. This is a serious meat thermometer: with a needle-tip probe, lightning-fast readings, and range up to 572°F (it can also be configured in Celsius), you can be both scientist and artist in the kitchen.

The tiny thermocouple sensor in the tip means that not only are the readings super-fast (about 3 seconds), you can check temperatures of even the thinnest of foodstuffs.

Plus, the company's customer service is great. A few months later, a handy info booklet showed up in the mail — again, out of the blue — but this time from ThermoWorks. The booklet is now included with all thermometers, so they sent them out to previous customers, too! Now that's taking care of business.

—Arwen O'Reilly Griffith

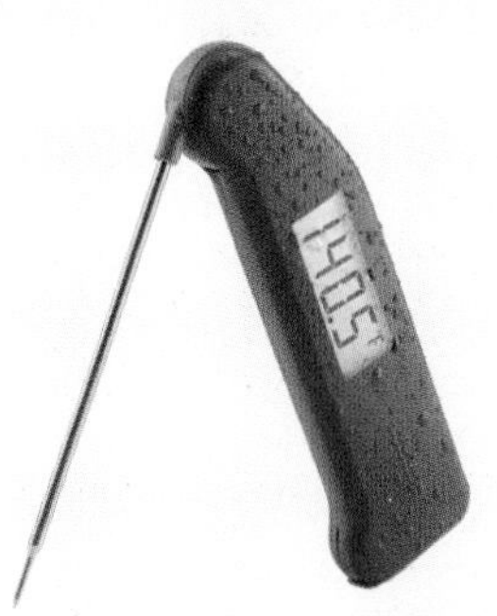

Pistols, Whipped

Badass Lego Guns
by Martin Hüdepohl
$30 No Starch Press

It's rare that a book's title so elegantly sums up its virtues, but with *Badass Lego Guns*, well, that's what you get. It's a manual for creating five different gun-shaped Lego models that fling Lego bricks and rubber bands instead of bullets.

Each entry begins with an explanation of the most important aspect of the gun: its firing mechanism. Next, to ensure that you can actually build the model, a parts list is provided. Finally, each gun gets full step-by-step CAD drawings similar to those found in regular Lego kits, except they're black and white with red accents highlighting the relevant Lego elements, which works quite well.

The author's models are cool and challenging, and the "Warbeast" model, featured on the cover, is particularly sweet. However, even the simpler models offer an excellent building challenge and, afterward, a fun toy.

—JB

Geeks on a Budget

Snip, Burn, Solder, Shred: Seriously Geeky Stuff to Make with Your Kids
by David Erik Nelson
$25 No Starch Press

Culled from the curriculum of what the author fondly calls "The Hippie School for Troubled Youths," these creative projects share a down-to-earth sensibility.

The 24 projects are broken down into three sections of roughly increasing difficulty, but all are fun, inspired, and accessible to a wide range of ages.

The premise is simple: for each project, you'll make something cool, do it cheaply (or for free), and pick up a transferable skill or fundamental understanding of the thing you made.

The skills involved range from sewing to basic carpentry and electronics, with every project explained from the most basic level. For your little rocker, there's an entire section of off-the-wall music projects, like the X-Ray Talking Drums, Electro-Didgeridoo, and Cigar-Box Synthesizer.

—Bruce Stewart

Give Yourself a Lift

***Moving Heavy Things* by Jan Adkins**
$14 Wooden Boat Publications

The title says it best. If you ever sit around imagining how to maneuver that gigantic welding project out of your shop, or how to lift something beyond your strength without injury, or heck, if you've ever wanted a good primer on being fiendishly clever in the world of simple machines, then look no further.

This charmingly illustrated ode to applied physics is surprisingly extensive for its 47 pages. Adkins succinctly manages to combine useful historical anecdotes and heirloom methods with charts, equations, and step-by-step instructions for lifting or scooting just about anything.

A delightful guide to getting the job done with back, joints, fingers, toes, and ego intact. *—Meara O'Reilly*

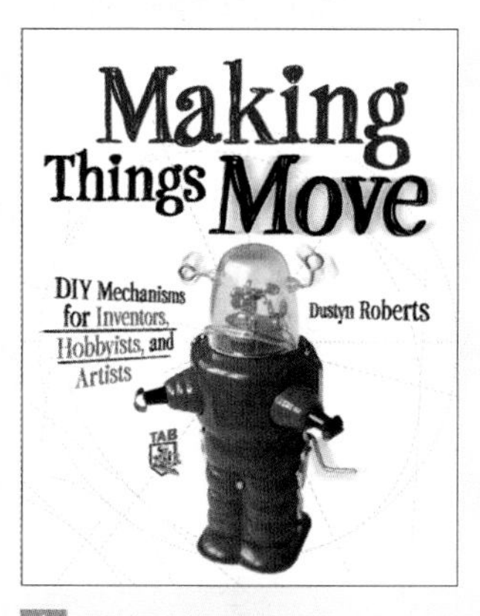

Move On

Making Things Move
by Dustyn Roberts
$30 McGraw-Hill

This book was written for me, by which I mean it was written for you — if you happen to be someone with ideas but no experience in, well, making things move.

It literally starts at the very beginning, patiently explaining basic (then less basic) physics, and moves through all the components of a successful project: different materials and what they're best at, the myriad ways of fastening parts together, all the different options for powering and controlling movement, etc.

If you're truly a newbie, you can do each mini "project" as you read through, or if you have experience in, say, drilling holes or breadboarding a circuit, you can skip them.

Since this is the 21st century, after all, the book finishes up with a breakdown of designing in 3D and having parts fabricated, but author Dustyn Roberts takes as much pleasure in hand-cranked mechanisms as in stepper motors.

She also has a rollicking good time ferreting out clever tricks and useful bits of information, like how to drill a centered hole minus a lathe.

Let's be honest — no one knows how to do everything, but *Making Things Move* does its best to get you there. *—AOG*

Mesh Monitoring

Building Wireless Sensor Networks
by Robert Faludi
$35 O'Reilly Media

If you want to add wireless communication to your electronics projects but don't know where to start, this book serves as the perfect guide.

Robert Faludi demystifies the confusing realm of ZigBee wireless mesh networking and walks you through basic examples that serve as great starting points for many fun wireless projects.

Building Wireless Sensor Networks starts with basic point-to-point serial communication using XBee radios so your microcontroller can communicate wirelessly with your computer or another microcontroller.

Faludi goes on to demonstrate how to wirelessly read the XBee radio's analog and digital I/O pins so you can connect sensors directly to your XBee without the need for a microcontroller. Not only can you learn to collect wireless sensor data, such as temperature, motion, or light, you can learn to create a basic home automation system to control appliances wirelessly.

Building your own wireless sensor network will be a much less daunting task with this book at your side. And if you're interested in getting into more advanced ZigBee mesh networking applications, it serves as a great starting point.

—Matt Richardson

Maybe it's the techno-utopian in me who thinks that the right tool, the right part, the right material, will make everything, well — all right. Apparently, I'm not the only one obsessed with catalogs; when we put out a query online, we got tons of suggestions (makezine.com/go/catalogs). Every maker who dabbles in electronics needs the Digi-Key (digikey.com), Jameco Electronics (jameco.com), and Newark (newark.com) catalogs. Tool freaks drool over Garrett Wade (garrettwade.com) and Lee Valley (leevalley.com). Besides these staples, here are some others you might want to check out.

—Gareth Branwyn

Electronic Goldmine

goldmine-elec.com

Dan Barlow of HacDC writes: "The Electronic Goldmine offers discounts on SMT components and often has cool, weird stuff like IC masks and wafers. They also have a special robot section."

MPJA

mpja.com

A bunch of makers recommended the venerable Marlin P. Jones & Associates catalog. MAKE's Collin Cunningham: "MPJA has awesome prices on a variety of electronics equipment. They also have dirt-cheap toggle switches and unusual surplus items, like a security cam mount I found useful for shooting macro-project builds."

Small Parts

smallparts.com

Small Parts doesn't have a print version anymore. But it's a great source for small orders of materials and supplies. Make: Online's Sean Michael Ragan writes: "This is where you go if you need to build, say, a gas chromatograph from scratch. "

All Electronics

allelectronics.com

A copy of the All Electronics catalog will likely be found on every wire-head's bookshelf. It's always been on mine. I haven't actually ordered anything from them in a while, but I always enjoy scanning their offerings of new and surplus parts.

Lindsay's Technical Books

lindsaybks.com

If every electronics nerd has the Digi-Key and Jameco catalogs on her/his bench, every steampunk and retro-techie covets this catalog. Sean Ragan: "Lindsay's features reprints of classic machine manuals and books on fundamental technology skills. A great depository of forgotten lore."

Plastruct

plastruct.com

The scale-model plastic stock company that launched a million train boards, architectural models, and low-budget sci-fi special effects shots. It offers everything at various scales: I-beams, T-beams, H-beams, tubing, tiny plumbing parts, stone and brick-patterned sheets, plastic sheet stock — you name it.

Micro-Mark

micromark.com

Sean Ragan writes: "Micro-Mark sells tools intended for model builders, including many tools that are custom-manufactured for this catalog. They offer casting supplies, small clamps, gluing jigs, micro machine tools, and more."

Polytek Plastics

polytek.com

MAKE and CRAFT's Becky Stern says: "Polytek not only has detailed descriptions of the products they sell but also loads of moldmaking advice (with pictures!). The catalog has information about people using their materials and the industries in which they work (props, special effects, landscaping, concrete, architectural work, etc.)."

Scenic Express

scenicexpress.com

Anybody working in miniatures, from dollhouse makers to wargamers, will find incredible stuff here. They carry all sorts of miniature trees, grass, foliage, and buildings; molds for making rocks, bridges, and stonework; little plastic people; and all the tools and supplies you need to create your own Lilliputian wonderlands.

Tricks of the Trade By Tim Lillis

Deter bike thieves.

Sick of no-good thieves stealing your bike parts? Here's a handy trick from the good folks at Mojo Bicycle in San Francisco to help deter them.

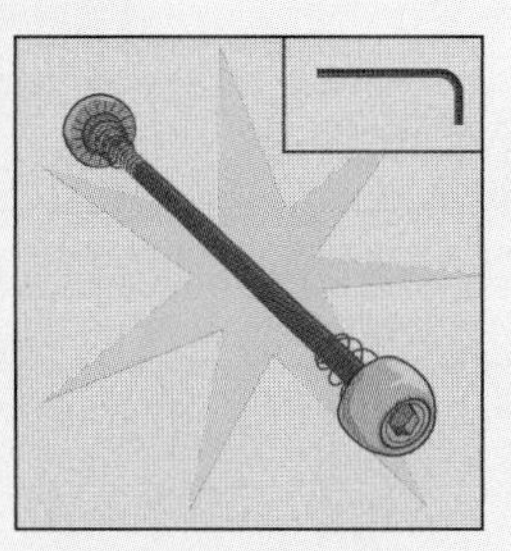

This particular trick involves getting a set of locking skewers for your wheels, the ones that use an Allen key to tighten and loosen.

Take some lip balm and fill most of the cavity in the Allen bolt. Insert a small ball bearing that fits the hole but is small enough to pry out. If possible, completely cover the ball.

You're now ready to thwart some casual thieves, even if they have an Allen wrench! Works great for seat posts too. You may want to carry something (in your kit, or hidden in your U-lock) to pry the ball out.

Have a trick of the trade? Send it to tricks@makezine.com.

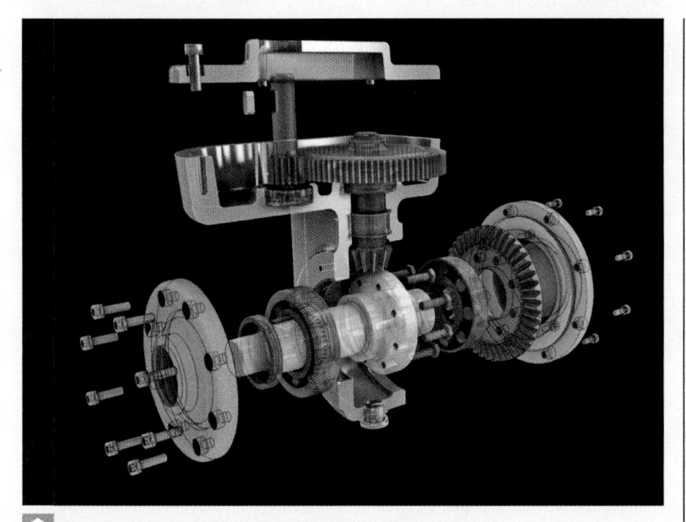

Autodesk Inventor Publisher 2011

$2,495 store.autodesk.com

Autodesk sent MAKE Labs a copy of its latest software, Inventor Publisher 2011, so I got to test its various features. The program comes with preset visual styles for wood, metal, plastics, and more, and is fully customizable. It also lets you insert callouts, arrows, and labels that help direct attention to specific details. These images can then be used to create technical manuals, user guides, and any other form of visual documentation, including videos.

But the true power and application of the software is realized when you export your document onto the Autodesk server. There, other people can access your files and explore your model and instructions using the free Inventor Publisher Mobile Viewer app. Download the app and check out the next step in digital documentation. *—Nick Raymond*

John Baichtal is a contributor to MAKE and the GeekDad blog on wired.com.

Marc de Vinck is director of product development for the Maker Shed.

Saul Griffith is a columnist for MAKE and chief cyclist at onyacycles.com.

John Edgar Park is a frequent contributor to Make: Online.

Tim Lillis is a freelance illustrator and DIYer.

Sean Michael Ragan is a jack-of-all-trades and master of, er — almost one.

Nick Raymond is one of MAKE's engineering interns.

Matt Richardson is a technophile, tinkerer, video producer for Make: Online, and co-host of Make: Live.

Max Rohde is 16 and currently attending the Alameda Science and Technology Institute.

Christopher Singleton, a father of three boys, is a maker, inventor, writer, and product development specialist.

Bruce Stewart is a freelance technology writer and contributor to O'Reilly's Radar blog. radar.oreilly.com

*** Want more? Check out our searchable online database of tips and tools at makezine.com/tnt.**

Have a tool worth keeping in your toolbox? Let us know at toolbox@makezine.com.

TOY INVENTOR'S NOTEBOOK

Make a Diving SpudMarine!

You can make it!

INVENTED AND DRAWN BY **Bob Knetzger**

HERE'S A FUN AND FAST WAY TO CREATE a homemade version of the classic cereal premium toy, the baking powder-powered diving submarine. Instead of plastic, we'll make one out of a potato. It works well — and when you're done, you can eat it!

First, cut a potato into a ¾"–1" diameter cylinder, about 3" long. Use a piece of ¼" brass tubing as a plug cutter. Press the tube all the way into the potato to make 3 through-holes evenly spaced along its length. This will reduce the mass of the potato for better diving. Enlarge the bottom of the center hole to make a flared conical opening (see #4). This will create an air chamber for the bubbles.

Next, make a periscope. Cut a thin piece of wood about ¼"×1" and drill a ¼" hole in the center. Cut a 1"-long piece of ¼" wood dowel and push it through the hole, so there's an equal amount of dowel exposed on each side. Insert the periscope into the middle hole in the potato.

Time to test it! Place the SpudMarine in a tall pitcher or vase of water. If it floats, cut off some of the periscope and try again. If it sinks fast, trim off some of the potato and try again. If it sinks very, very slowly, it's ready! The sub should be just ever so slightly heavier than neutral buoyancy for best diving action.

Remove the sub and shake it dry. Pack some baking powder (not baking soda) into the bottom of the center hole. Use another piece of wood dowel to tamp it in tightly. Gently lower the sub back into the water and let it sink. Then watch it bubble, rise, breach the surface and "burp" its bubbles — and sink again! How many cycles will your sub do before reloading the powder?

Bob Knetzger is an inventor/designer with 30 years' experience making all kinds of toys and other fun stuff.

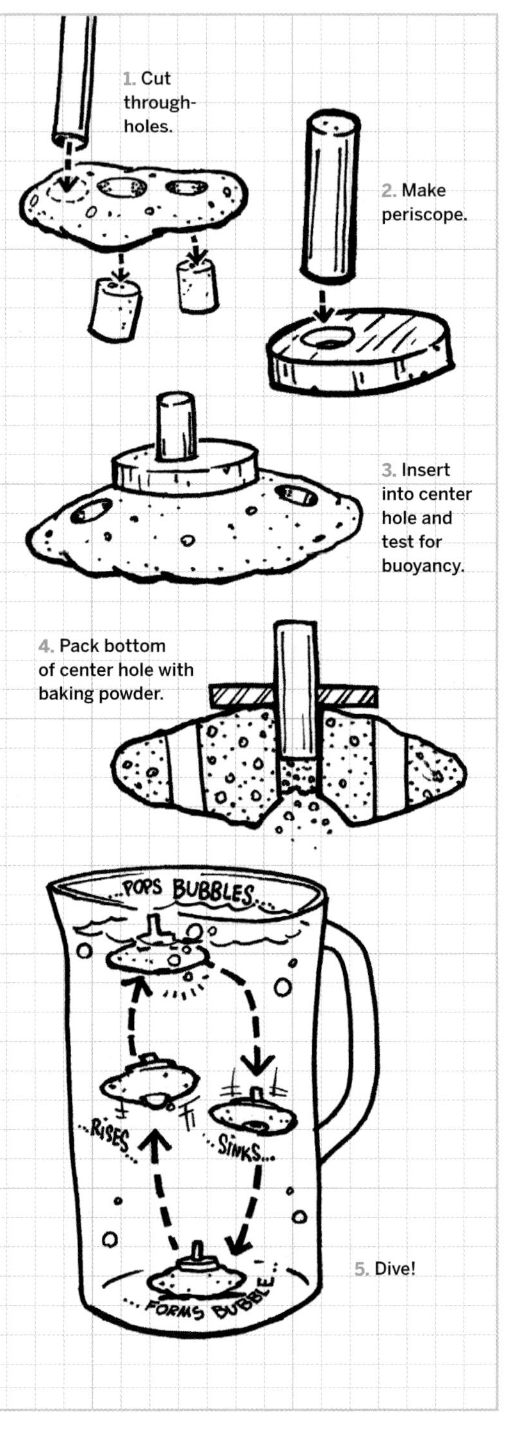

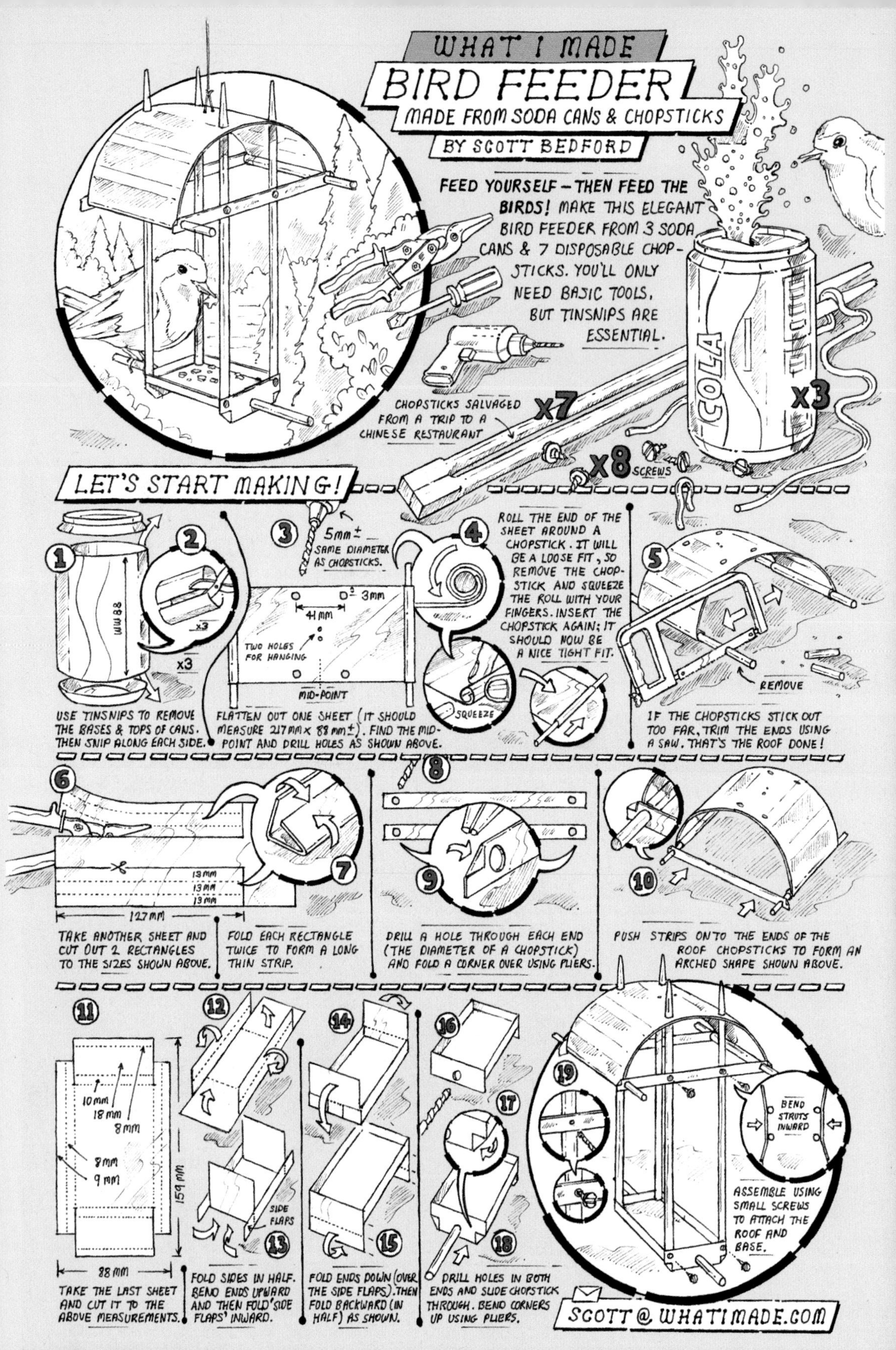

WHAT I MADE
BIRD FEEDER
MADE FROM SODA CANS & CHOPSTICKS
BY SCOTT BEDFORD
FEED YOURSELF – THEN FEED THE BIRDS! MAKE THIS ELEGANT BIRD FEEDER FROM 3 SODA CANS & 7 DISPOSABLE CHOPSTICKS. YOU'LL ONLY NEED BASIC TOOLS, BUT TINSNIPS ARE ESSENTIAL.
COLA
x3
x7
CHOPSTICKS SALVAGED FROM A TRIP TO A CHINESE RESTAURANT
x8 SCREWS
LET'S START MAKING!
1
2
88mm
x3
x3
USE TINSNIPS TO REMOVE THE BASES & TOPS OF CANS. THEN SNIP ALONG EACH SIDE.
3
5mm ± SAME DIAMETER AS CHOPSTICKS.
3mm
41mm
TWO HOLES FOR HANGING
MID-POINT
FLATTEN OUT ONE SHEET (IT SHOULD MEASURE 217mm x 88mm ±). FIND THE MID-POINT AND DRILL HOLES AS SHOWN ABOVE.
4
ROLL THE END OF THE SHEET AROUND A CHOPSTICK. IT WILL BE A LOOSE FIT, SO REMOVE THE CHOPSTICK AND SQUEEZE THE ROLL WITH YOUR FINGERS. INSERT THE CHOPSTICK AGAIN; IT SHOULD NOW BE A NICE TIGHT FIT.
SQUEEZE
5
REMOVE
IF THE CHOPSTICKS STICK OUT TOO FAR, TRIM THE ENDS USING A SAW. THAT'S THE ROOF DONE!
6
13mm
13mm
13mm
127mm
7
TAKE ANOTHER SHEET AND CUT OUT 2 RECTANGLES TO THE SIZES SHOWN ABOVE.
FOLD EACH RECTANGLE TWICE TO FORM A LONG THIN STRIP.
8
9
DRILL A HOLE THROUGH EACH END (THE DIAMETER OF A CHOPSTICK) AND FOLD A CORNER OVER USING PLIERS.
10
PUSH STRIPS ONTO THE ENDS OF THE ROOF CHOPSTICKS TO FORM AN ARCHED SHAPE SHOWN ABOVE.
11
10mm
18mm
8mm
8mm
9mm
159mm
88mm
TAKE THE LAST SHEET AND CUT IT TO THE ABOVE MEASUREMENTS.
12
13
SIDE FLAPS
FOLD SIDES IN HALF. BEND ENDS UPWARD AND THEN FOLD 'SIDE FLAPS' INWARD.
14
15
FOLD ENDS DOWN (OVER THE SIDE FLAPS). THEN FOLD BACKWARD (IN HALF) AS SHOWN.
16
17
18
DRILL HOLES IN BOTH ENDS AND SLIDE CHOPSTICK THROUGH. BEND CORNERS UP USING PLIERS.
19
BEND STRUTS INWARD
ASSEMBLE USING SMALL SCREWS TO ATTACH THE ROOF AND BASE.
SCOTT@WHATIMADE.COM

HEIRLOOM TECHNOLOGY
By Tim Anderson

Chinampa Gardens

Can ancient Aztec agriculture save the sinking Sacramento Delta islands?

A KATRINA-STYLE DISASTER IS BREWING in California. Halfway between Sacramento and San Francisco there's a vast labyrinth of large islands in the Sacramento-San Joaquin River Delta. It's rich cropland, a popular recreation area, and home to suburban neighborhoods.

Unfortunately the islands are sinking, as fast as 3 inches a year. Levees hold back the water, but some fields are more than 25 feet below sea level. All the islands have flooded at least once. When that happens, thousands of acres are inundated, and saltwater flows upstream from San Francisco Bay, tainting the freshwater Delta, source of drinking water for 22 million people.

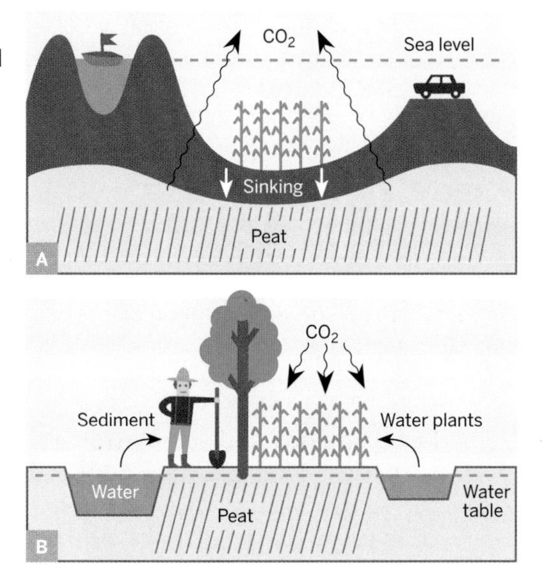

The Delta Islands Are Sinking

Figure A shows a Delta island that's pumped dry for agriculture. Oxygen diffuses into the drained soil; this allows microorganisms to consume the subsurface peat and exhale carbon dioxide (CO_2) into the atmosphere.

As the peat disappears, the island collapses. It's amazing and confusing to be driving along, look up, and see a boat going by at the top of a levee. The bottom of that channel is higher than your head!

On flooded islands, the marsh plants don't regrow because the water is now too deep. The U.S. Geological Survey's paper "Delta Subsidence in California" suggests a solution: "Shallow flooding to mitigate subsidence by slowing peat oxidation and allowing growth of wetland vegetation that contributes biomass accumulation."

That sounds a lot like the wetland gardens of the ancient Aztecs, called *chinampas*. This method of wetland agriculture was the basis of the Aztec civilization. Chinampas not only prevent land from sinking, they build soil and gradually *increase* the height of the fields.

Chinampa: Rising Soil and Fertility

Aztec chinampas maintained high fertility and produced multiple crops per year to feed the capital city, Tenochtitlan — one of the largest cities in the world when Hernán Cortés first saw it in 1519.

A high water table is maintained under the chinampa (Figure B). This prevents oxygen from reaching the underground peat and decaying it. Water wicks up to the roots of the crops from below. The farmer fertilizes the garden with aquatic plants and mud from the

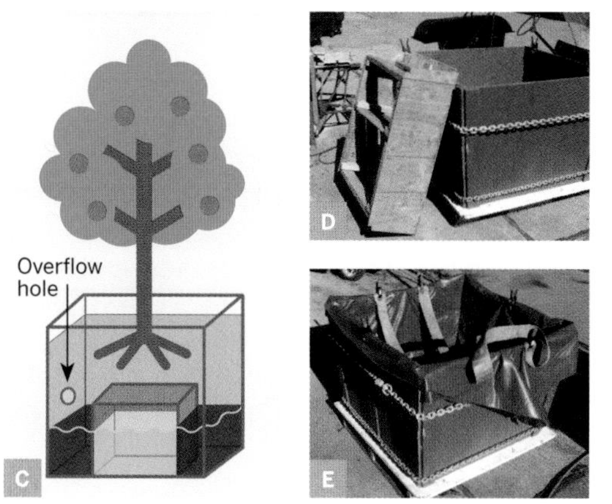

canals, building the soil up and capturing CO_2.

The tradition is still strong in places like Mexico City's popular Xochimilco flower gardens. Unfortunately the Spanish drained Tenochtitlan's lakes for pastures and urban sprawl; the canal patterns of former chinampas are still visible in fields and streets.

MATERIALS AND TOOLS

Lumber, plywood, and wood screws to build a planter box and internal soil platform
Waterproof tarp or tub for a water tank
Plants and potting soil
Pipe two lengths, about the same depth as your planter
Bottle, yogurt cup, scrap of screen

Experiment with Wetland Agriculture

Here's how to make a self-watering planter that works a lot like a chinampa.

→ START

1. Build a sturdy box and soil platform.

The soil platform is 1' narrower than the box (Figures C and D), so the dirt extends down into the water along 2 edges. This waterlogged dirt wicks water up to the rest of the soil. Another method is to hang rags down through the platform to wick water up.

I made my soil boxes deep so I could plant a mature fruit tree in 2' of soil and 1' of water. If you're just planting vegetables, 6" of soil is supposed to be enough. Heavy mulch helps control weeds and reduces evaporation.

2. Add the water tank.

I lined my box with a heavy tarp to make the water tank (Figure E), then placed the wooden platform inside. (Those heavy straps let me haul out the platform if it gets rotten.)

3. Add the refill pipe and overflow port.

The overflow hole prevents the water level from rising too high and drowning the plants' roots (Figure F). Cover it with a bit of screen, or mosquitoes will breed in the tank. A little overflow is good; it will prevent salt buildup in the soil and the water tank.

The white yogurt cup covers the water refill pipe, which extends down into the tank.

4. Add soil and plants.

Any soil mix that's good for potting or gardening is fine. I relocated a mature orange tree into this planter, and planted tomatoes and other veggies all around it. After a few weeks, the tree was sprouting new leaves like crazy.

5. Add a refill indicator.

A tube sealed to the mouth of this inverted bottle extends to the bottom of the tank (Figure G). When the tank is empty, air can enter the tube, and the bottle empties.

I'm amazed at how little work my chinampa garden was. I went away for two weeks and it tended itself. My friends' garden beds died when they forgot to water for a few days. My vines were still making flowers and tomatoes in February.

Tim Anderson (mit.edu/robot) is the co-founder of Z Corp. See a hundred more of his projects at instructables.com.

Photography by Tim Anderson

REMAKING HISTORY
By William Gurstelle, workshop warrior

Ptah-Hotep and the Bag Press

Use linen, wood, and the wisdom of ancient Egyptians to make a juicer.

THE ANCIENT EGYPTIAN RUINS AT Memphis stretch nearly 20 miles along the banks of the Nile River. The oldest artifacts date back more than 4,500 years to the Old Kingdom, the period during which the first pyramids were built. At the spot called Saqqara on the river's western shore is a particularly spectacular area of temples, tributes, and tombs, achievements credited to the laborers working for the pharaohs of the Third, Fourth, and Fifth Dynasties.

Among the pyramids and sphinxes lies the mastaba (a square, bench-like tomb) of Ptah-Hotep, the prime vizier and close confidant of Fifth Dynasty pharaoh Djedkare Isesi. The size and grandiosity of Ptah-Hotep's final resting place are evidence that he was a BMON (big man on the Nile) and a person of great wealth and influence.

The scene on the wall of Ptah-Hotep's tomb particularly interests archaeologists because it shows in great detail the state of Egyptian industry and technology at the time of his demise (right). There are clear depictions of early boat building, cloth weaving, fish drying, and jewelry making. Perhaps the most interesting is the detailed illustration of the oldest form of food processing equipment: the Egyptian bag press.

Some clever Egyptian thinker, perhaps even Ptah-Hotep himself, discovered how much more juice could be extracted from grapes, dates, or olives by wrapping the fruit in a linen bag and then twisting the ends. Like wringing water from a towel, this technique exerts torsional forces on both ends of the bag that apply enormous pressure to whatever is inside. The press wrings out more juice than would be possible by other methods, including smashing grapes with mallets or feet.

TEAM EFFORT: As seen in these hieroglyphs, the original bag press design employed five men working together, twisting and separating the poles.

Archaeologists call the operation of bag presses "wringing the cloth." Images of the process appear frequently in hieroglyphs found throughout Egypt. In the Ptah-Hotep hieroglyph, five men working together use a linen bag and poles to extract juice from grapes. Two men twist each stick in opposite directions while a fifth has the unenviable job of using his body to separate the two ends as the contracting motion of the twisting bag draws them together.

Three hundred years later, an improved version of the bag press appears in the hieroglyphic record. The ends of the bag are placed through holes in large wooden separators

Image Corbis

Like wringing water from a towel, exerting torsional forces on both ends of the bag applies enormous pressure to whatever is inside.

and the twist is applied outboard to these, making the process easier and more efficient. One might wonder why it took the Egyptians so long to come up with the idea of a simple frame to hold the bag ends apart, but perhaps the pace of technology was slower back then.

MATERIALS

Wood boards, 2×6: 12" long (1), 6½" long (2)
Nominal 2×6 dimensional lumber actually measures 1½"×5½".
Plywood, 11"×8", cut into four 8"×5½" triangles as shown in the frame diagram (Figure C)
Deck screws: 2½" long (6), 1¼" long (24)
Iron pipe flanges, ¾" NPT (2)
Wood screws, #12×1" (8)
Linen cloth, approximately 12"×24"
Hardwood dowels, ⅝" diameter, 8" long (2)
Glass loaf pan, approximately 4½"×8½"×3"

NOTE: To make the bag press as authentic as possible, you can use cedar (a wood available to early Egyptians) for the 2×6 pieces and substitute thicker, solid cedar pieces for the plywood bracing triangles. Obviously, the ancients didn't have steel deck screws. They relied on mortise-and-tenon joints and pegs. If authenticity is important, you can use joinery techniques instead; check with a local woodworking expert for assistance on joint making.

TOOLS

Drill and 7⁄64" bit
Spade bit, 1"
Driver bits for deck screws
Sewing machine, or needle and strong thread

→ START
Build an Ancient Egyptian Bag Press

1. Prepare the bag.

Begin by laundering the linen cloth. Fold one short end twice over and stitch a hem, using a sewing machine or sewing by hand with very fine stitching and strong thread. Hem the other end the same way.

Fold the fabric in half lengthwise with the hems facing out. Stitch along the long edge (Figure A).

Turn the bag inside out so the hems and the raw edges of the long seam are hidden inside the bag. Fold over one open end to make a sleeve large enough to accommodate the diameter of the turning rods.

Sew the sleeve, leaving its top and bottom open for the rod. Leave the other end of the bag open (Figure B).

2. Prepare the uprights.

Locate and mark the holes in the uprights according to the frame diagram (Figure C, following page). Drill a 1" hole in each upright.

3. Build the frame.

Use the long deck screws to make a U-shaped

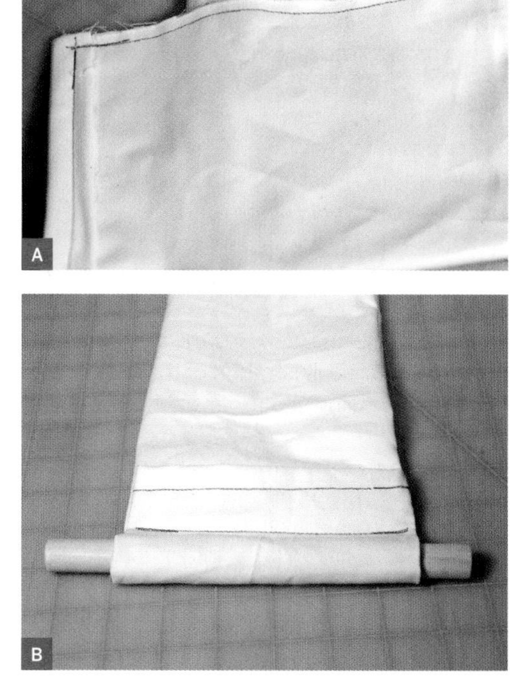
A
B

Photography by Sam Murphy

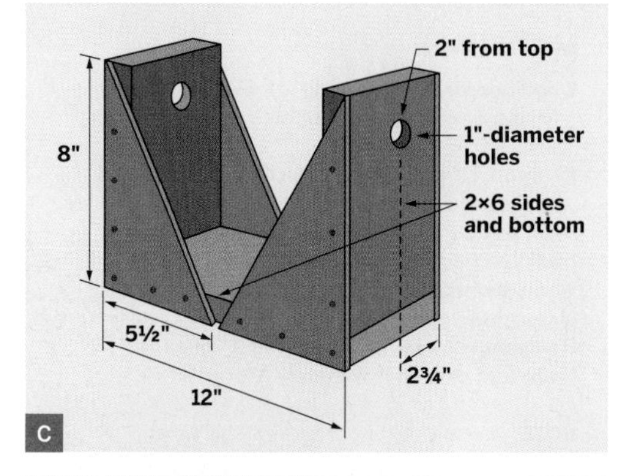

C

D

E

F

frame as shown. Secure each 6½" length of 2×6 with 3 long deck screws through the 12" length of 2×6. Fully seat the deck screws into the wood.

Attach the bracing triangles to the frame using the shorter deck screws (Figure D). Test the frame for rigidity.

Center the pipe flanges over the holes in the upright, mark and drill pilot holes with the 7⁄64" bit, then attach the flanges using the short #12 wood screws (Figure E).

4. Operate the bag press.
Place the loaf pan on the frame.

Insert a small quantity of soft, juicy fruit, such as grapes or peeled orange sections, in the bag. Insert the ends of the bag through the holes in the uprights.

Insert one of the dowels through the sleeve, centering the dowel on the flange hole. Tie the open end of the bag around the middle of the other dowel (Figure F).

As you rotate the dowels in opposite directions, the bag will tighten, compressing the fruit, expressing the juice into the pan, and leaving the pulp in the bag.

William Gurstelle is a contributing editor of MAKE. His new book, *The Practical Pyromaniac*, will be available in June 2011 in the Maker Shed (makershed.com) and at other fine booksellers.

Extended Paintbrush
To paint high places from ground level, glue a paintbrush to a lightweight pole. It's a little slower painting this way, but I feel much safer not trying to climb 20 feet up. I've found that Gorilla Glue or a construction-grade glue is the best for holding the brush on. —*Milton Ammel*

Find more tools-n-tips at makezine.com/tnt.

Join other Makers on a
Geek Canoe Trip
5 relaxing days on Canada's Peace River
Gadgets, inventions, and challenges!
An unforgettable vacation
See www.flownorth.ca/make for the details
Flow North
Paddling Company
Phone: 1-877-926-2649
Skype: flow-north-paddling
info@flownorth.ca

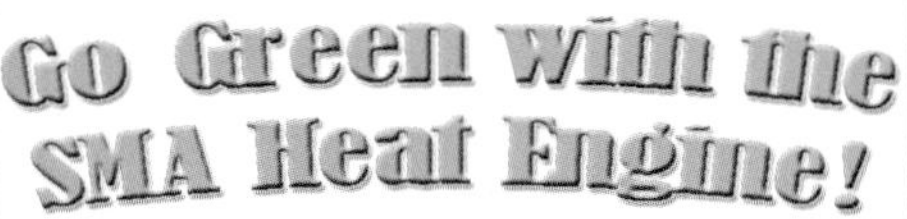
Go Green with the
SMA Heat Engine!

Cold water
Hot water

MUSCLEWIRES.COM

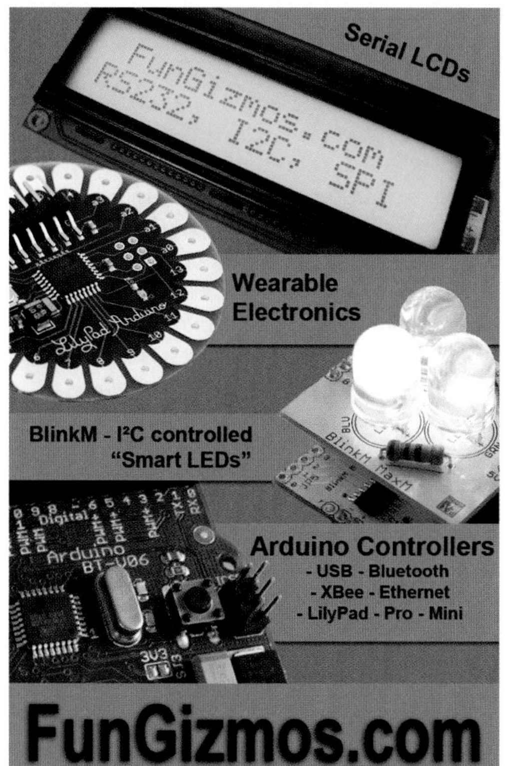
Serial LCDs
FunGizmos.com
RS232, I2C, SPI
Wearable
Electronics
BlinkM - I²C controlled
"Smart LEDs"
Arduino Controllers
- USB - Bluetooth
- XBee - Ethernet
- LilyPad - Pro - Mini
FunGizmos.com

iFixit
Parts, Upgrades
Tools & Guides
Love quality tools?
We sure do.
DIY
Must Have!
54 Piece
Bit Driver Kit
$19.95
Includes a magnetized driver, flexible extension, and bits to open just about anything.
Find what you're looking for at iFixit.com

HOMEBREW

My Home-Built Power Wheelchair By Marcus Brooks

ABOUT THREE YEARS AGO I ACCEPTED arthritic degeneration of my hucklebone (hips) as an excuse to build a power wheelchair. It's a dangerous project that could injure or kill me if I'm not careful. Several components, from the joystick MPU to the rear casters, are used in ways forbidden by manufacturers' fine print, so follow my example at your own risk!

The chair uses two 500W electric scooter motors to drive wheels meant for a go-kart. Tractor-lug wheelbarrow tires would fit (and be awesome), but non-marking wheelchair tires are more politic. A jackshaft adapts between #25 chain for the motors and #35 for the wheels, and allows gearing adjustment up to about 30:1. McMaster-Carr saved the project with #25 sprockets stocked in the right size. The motors and jackshafts are on a sliding sub-frame for independent tensioning of the chain stages.

The motors are powered by two U1-size AGM batteries via a Robot Power Sidewinder speed controller. A Zilog Z8 EncoreXP microcontroller converts the analog output of a game joystick to radio-control-style servo pulses for the Sidewinder. Fail-safe control is vital, so code and test carefully! PIC and Atmel microcontrollers seem more popular, but I'm familiar with Zilog chips, and they're cheap. At about 10 MIPS, my joystick's MPU is as powerful as the CDC 6600 mainframe I used in school!

The chair frame is 14-gauge square carbon steel tubing, which I cut and joined with a 4" angle grinder and 110V flux-core wire welder. (Flux-cored welding is easy enough, but takes some practice; some of my first welds failed.)

The most eye-catching feature of my chair is its seat. Camouflage bass boat seats are just cheaper in my size than wheelchair seats! I added angle steel extensions to the back to make the seat a little deeper. Zipper pouches in the same camo pattern were a lucky find.

Marcus Brooks is a technical writer in Austin, Texas.

Photograph by Marcus Brooks